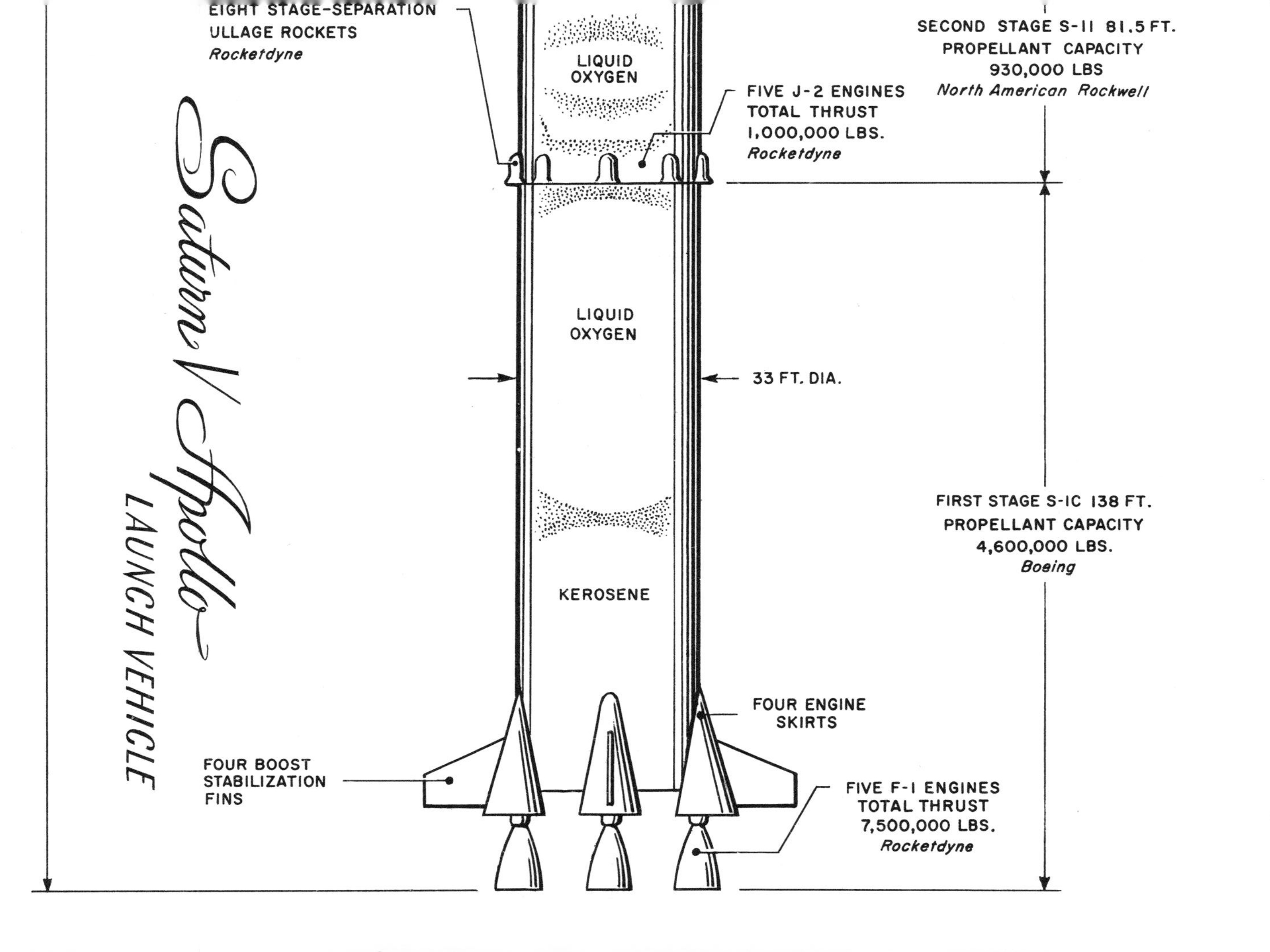
SECOND STAGE S-II 81.5 FT.
PROPELLANT CAPACITY
930,000 LBS
North American Rockwell
EIGHT STAGE-SEPARATION
ULLAGE ROCKETS
Rocketdyne
LIQUID OXYGEN
FIVE J-2 ENGINES
TOTAL THRUST
1,000,000 LBS.
Rocketdyne
LIQUID OXYGEN
33 FT. DIA.
FIRST STAGE S-IC 138 FT.
PROPELLANT CAPACITY
4,600,000 LBS.
Boeing
KEROSENE
FOUR ENGINE
SKIRTS
FOUR BOOST
STABILIZATION
FINS
FIVE F-1 ENGINES
TOTAL THRUST
7,500,000 LBS.
Rocketdyne
Saturn V Apollo
LAUNCH VEHICLE

EARTHBOUND ASTRONAUTS

THE BUILDERS OF APOLLO-SATURN

Previous Books by Beirne Lay, Jr.

I Wanted Wings
I've Had It
Twelve O'clock High
(in collaboration with Sy Bartlett)
Someone Has to Make It Happen

EARTHBOUND ASTRONAUTS

THE BUILDERS OF APOLLO-SATURN

BEIRNE LAY, JR.

PRENTICE-HALL, Inc., Englewood Cliffs, N.J.

Earthbound Astronauts: The Builders of Apollo-Saturn
by Beirne Lay, Jr.

ISBN 0-13-222307-4 (casebound)
ISBN 0-13-222331-7 (paperbound)
Library of Congress Catalog Card Number: 78-145628
Printed in the United States of America *T*

Prentice-Hall International, Inc., London
Prentice-Hall of Australia, Pty, Ltd., Sydney
Prentice-Hall of Canada, Ltd., Toronto
Prentice-Hall of India Private Ltd., New Delhi
Prentice-Hall of Japan, Inc., Tokyo

PREFACE

Psychologists warn writers and speakers against "apologizing"—because right off one is placed in a position of dealing from weakness. But if not an apology, at least an explanation ought to be given to those individuals and companies, more than worthy of mention, whose names do not appear in these pages about the builders of the Apollo-Saturn "Stack."

The omissions were inevitable for any writer who did not have ten or more years at his disposal; I therefore limited myself to a few representatives of the kind of men whose combined efforts on Apollo-Saturn bode well for the far larger programs that our nation is now mounting to preserve the habitability of the Earth, and perhaps ensure our very survival. The essence of the Apollo-Saturn achievement has been teamwork on a stupendous scale under a management philosophy that emphasized *communication* between the participants to an unprecedented degree. It has been said, partly in jest, that "when the weight of the paperwork equals the weight of the Stack, then you are ready to launch." There was no other way.

Every man I interviewed whetted my appetite for more such encounters and sharpened the keenness of my regret that time would not permit me to sit down with everyone who contributed significantly to the device that carried astronauts moonward. My best hope is that the ones omitted will agree that those mentioned may fairly represent them by proxy.

CONTENTS

EARTHBOUND ASTRONAUTS

THE BUILDERS OF APOLLO-SATURN

THE FOUNDATION

"You can't get there from here."
Rural colloquialism

No more hair-raising decision ever confronted a test pilot. He had less than three seconds in which to make it, while an audience of millions watched on TV, when the engines of his Gemini 6 spacecraft abruptly shut down almost at the instant of liftoff on one of the flights preliminary to the Apollo series.

Walter M. Schirra, command pilot, had the choice of pressing the panic button that would trigger the escape system or of sitting tight and keeping his powder dry on the assumption that the bird, pregnant with fuel that had been racing through the pumps, would not blow up. He sat tight. Nothing happened. Mission merely postponed.

Schirra's cool judgment was rightly recognized and applauded. What was not widely known was that two other men, ground controllers without the prestigious background of a veteran Schirra, were confronted by the same challenge, with comparable authority and responsibility. The remarkable aspect of this episode is that all *three* men arrived simultaneously at the right decision.

How? Because they were not guessing. They *knew.* (An electrical malfunction had caused the shutdown.) They knew because just such a contingency had been foreseen. And it had been foreseen because of the dedication of the engineers, scientists, and manufacturers of the Mercury, Gemini, and Apollo programs to the philosophy that every possible knowable source of failure must be either eliminated, backed up by an alternate system, or so well understood as to minimize or nullify its potential hazard.

Stated more succinctly, the creed has been, "Thou shalt have no other god but a tested and proven fact." Without it, and the manner of men who lived by it, and of whom I am writing this book, no American astronauts would yet have set foot on the Moon.

"Ignorance ain't what's wrong with the world," Will Rogers once said. "Everybody knows plenty. It's what they know that *ain't so* that causes the trouble."

The cliché that engineers and their ilk are dull fellows is a fitting target for Rogers' pithy shaft. After months of interviews and travel, I now know that the men who conceived and built Apollo-Saturn (Saturn consisting of the three rocket stages that boost the "payload" to departure speed from Earth orbit, and Apollo consisting of the manned payload that continues on to the Moon) are far indeed from being dull fellows or conforming to a mold. It is precisely because these ingrained skeptics are the natural enemies of what "ain't so" that they become dramatic, locked as they are in unending and—surely in the case of Apollo—mortal combat.

Dullness is commonly equated with the absence of human fireworks, whereas emotional conflicts are the fuel of drama. Yet paradoxically, the most dramatic achievement in history, the construction and voyage of Apollo 11, was made possible by that manner of man who can purge the last vestige of emotion from his performance, in the traditional scientific approach. The emotions were there, all right. Profound emotions. Yet the men I met, playing in a real spectacle transcending any ever seen on a stage, had been able to rise to a degree of objectivity bordering on the supernatural.

Norman Mailer, writing for *Life* magazine, wrestled turgidly with the question of whether the Apollo program was the work of God or the Devil. His forebodings were not shared by the builders of Apollo-Saturn whom I encountered. These were men who had banded together in a congregation that attends the same church in worship before a common god: a *Fact.* And when they knelt in awe before *any* fact, they did not blaspheme; nor are they necessarily out of harmony with their brethren who in other cathedrals worship a Savior who said, "I am the truth. And the truth shall set you free."

Thomas O. Paine, who succeeded James E. Webb as administrator of NASA, propounded a truth in reply to the monotonously recurring question: "Couldn't all that Apollo money have been better spent solving problems here on Earth?"

Paine replied in essence that any one of our major social problems is of so much greater a magnitude than landing men on the Moon that a sum like $24 billion would scarcely leave a dent. He could have cited the example of mental illness, which alone is costing this country $20 billion *per year*—nearly all that has been spent on Apollo in a decade.

Solutions for mental illness, poverty, racial strife, overpopulation, urban crises, and other human failures have historically been polluted by massive injections of damaging emotions. Paine expressed the hope that out of the Apollo experience might come new insights into man himself, the root of our afflictions.

Still struggling as we are in the Dark Ages of an understanding of human motivations (including our motivation for sending men to the Moon), we can learn at least two lessons from our Space program—the miracle that can result from the *un*emotional approach to a massive problem, and the paramount importance of proceeding step by step instead of plunging, as we have, into remedies for some of our social ills. Man has been given fresh grounds for confidence that he may yet achieve other "impossibilities," even create an environment on Earth of dignity for all men, if only he can learn, as has the Space pioneer, to keep his cool.

Amid the doubts and confusion and growing pains of the Apollo adventure, the leaders of NASA under Jim Webb found and kept their cool. In industry, veteran leaders like Lee Atwood of North American Rockwell, who headed the single company most deeply involved, likewise kept his cool. And so did the responsible heads and engineers of other major contractors such as Boeing, McDonnell Douglas, Grumman, IBM, Lockheed, GE, Garrett, Northrop, Aerojet, TRW, M.I.T., and their legions of subcontractors and vendors.

Poised for liftoff on Pad 39 at Cape Kennedy, the Apollo-Saturn vehicle resembled an enormous machine gun bullet standing 36 stories high, wrought of metal and electronics and high energy fuels. But that awesome "Stack" bespeaks not so much the presence of physical materials as the concentration of millions of facts, verified many millions of times. Nowhere throughout the Stack's 363 feet nor among its 6.5 million pounds, as the countdown ticked away on July 16, 1969, did there lurk a known hostage to what "ain't so" that could conceivably send Armstrong, Collins, and Aldrin to their doom. The thrust which sent the Stack onward to the Moon was the thrust of truth, the result of trial and error, of unglamorous research and development, and—if not of blood—at least of sweat and tears.

ii

It is of the modern-day men to whom fact is a religion, then, men as exciting in the fruits of their handicraft as poets, that I feel encouraged to write—men who in our age of technological explosion have added a new dimension to the inquiring mind of a Copernicus, a Galileo, a

Bacon or a Newton or an Einstein, the dimension being the "team" rather than the "solo" approach of the classical lonely genius.

My primary concern will be with that preponderance of the iceberg which lay below the surface, the anonymous heroes rather than the more publicized astronauts and the Apollo missions—the earthbound explorers whose imagination reached farther than astronauts have thus far traveled into the cosmos.

Frank Borman, commander of Apollo 8 on its pioneering voyage around the Moon on Christmas of 1968, was being as factual as he was eloquent when he declared: "We rode there on the shoulders of giants."

iii

The long road to Pad 39 was paved with nonfacts.

"The Earth is flat."

"The sun and the planets revolve around the Earth."

"The Earth is the center of the universe."

The earlier of the shoulders to which Borman referred belonged to the Babylonian, Egyptian, Greek, and Roman astronomers who discovered and asserted, often like Galileo at the risk of their lives, what the facts were.

"There's no way to overcome the force of gravity."

More recently, in the eighteenth century, an intellectual giant named Sir Isaac Newton spelled out the Universal Laws of Motion, defined gravity and orbital forces, and, in his Third Law of Motion, opened the highway to manned flight in Space in a single, simple sentence:

"For every action there is an equal and opposite reaction."

When its full implications were finally grasped, this truth was to serve as the principle of the rocket engine, without which travel in a vacuum would never have become possible. When the astronauts arrived, Newton's monument was already there on the Moon, as was that of Johannes Kepler, from whom Newton learned the laws of planetary motion.

"Humans cannot survive in the hard vacuum of Space."

True but irrelevant. Men needn't try to. They would have to learn how to take their environment along with them.

"Humans won't be able to function *in the absence of gravity."*

"Deprived of orientation out in Space, with no up or down, humans will become hopelessly confused, psychologically disturbed, physically ill."

"The odds against survival are too formidable. Boosters will blow up during launch, engines won't restart in a vacuum, meteorites and radiation belts will be lethal, capsules will burn up like shooting stars upon atmospheric entry."

Incredibly, Space travel as of this writing has been actuarially the safest mode of transportation ever invented—zero in-flight injuries or fatalities in the NASA program after millions of passenger miles at hypersonic speeds.

"The necessary reliability and precision in guidance and navigation are beyond attainment."

A red flag to a gyro-minded "bull" at M.I.T. named "Doc" Draper.

The foregoing highlights only a few of the misconceptions that were harbored by many experts of distinction up to the time the goal of a manned lunar landing was given serious support in 1958. As recently before that as 1953, a man who was to become a leader in the Apollo program, the late Dr. Hugh L. Dryden, confessed his misgivings:

"I am reasonably sure that travel to the Moon will not occur during my lifetime."*

Dryden's pessimism no doubt stemmed in part from the bleak outlook at that time for sufficient appropriations from Congress. But even James Webb, when he

*Although he died prematurely in 1964 and did not live to see the event, it did occur within his probable lifetime.

took over direction of NASA and its Apollo program in 1961, echoed Dryden's reservations when he spoke of "important areas of unpredictability, areas where none of the experiences of man could tell us what would be needed or how it could be provided."

iv

Fortunately for NASA, large areas of ignorance in rocketry and ballistics had been cleared before and during World Wars I and II by giants of three different nationalities: Konstantin E. Tsiolkovsky of Russia, Robert H. Goddard of the United States, and two Germans, Hermann Oberth (born in Hungary) and Wernher Von Braun.

The accolade "Father of Rocketry" has been most generally associated in this country with either Goddard or, more recently, Von Braun. But the latter himself gives precedence to Tsiolkovsky for being the first to relate rockets to astronautical theory and to perceive the need for multistaging, to Goddard for implementing theory into hardware through his experimental rockets beginning in 1915, and finally to Oberth, a younger man than Goddard, for his contributions between the two World Wars. Following chronologically comes Von Braun, who attained prominence in World War II when he shared with his superior, Dr. Walter Dornberger, the responsibility for developing the V-2 rockets which were launched against England.

Of the quartet, Goddard and Von Braun were the practical engineers, differing from each other radically, however, in that Goddard was a loner, suspicious of publicity and potential competitors, whereas Von Braun was a team player exceptionally gifted in spreading the gospel of Space exploration to the nonbelievers (Oberth was also a prolific writer and publicizer). From the first, all four recognized in the rocket the magic key to the door of Space.

In a vast oversimplification, but in the interests of

perspective, it can be said that the dream of landing a man on the Moon in the decade 1961-1970 owes its realization mainly to three pieces of hardware. The rocket engine. The gyroscope. The electronic computer. Omit any of the three and the dream tumbles over like a three-legged stool with two legs. All other hardware is secondary, even the Command Module in which the astronauts made their home—admittedly a vehicle of fantastic ingenuity and complexity. But it could never have been designed, checked out, and operated without the computer; it could never have reached its destination without the computer and the gyro, heart of the inertial guidance and navigation system, nor would it have lifted one inch off the pad without rocket engines of such concentrated might that comparisons to the power of all the hydroelectric dams in America harnessed together merely boggle the mind.

Another oversimplification: if one were to single out the least likely "miracle" of our manned mission to the Moon, it has to be the *decision*—the resolve to attempt it, and in only a decade. As a voter, I vigorously opposed both John F. Kennedy and Lyndon B. Johnson, but as a writer I find the conclusion to be inescapable that Johnson, as Chairman of the Space Council while he was Vice-President, laid the policy foundation for the momentous decision so imaginatively and fearlessly embraced by Kennedy. Here in all fairness were two pairs of the shoulders on which Borman's crew rode to immortality.

Also from the political rather than the scientific community were the shoulders belonging to Dr. T. Keith Glennan, the often forgotten first administrator of NASA from 1958 to 1961, who formulated many of the major decisions which culminated in Apollo's success, and his successor, James E. Webb, architect of the largest-scale managerial effort our country has witnessed outside of wartime, who, if he had done nothing else, would have to be remembered gratefully for a masterful job of extracting the needed money from a succession of Congresses.

A third, and I hope pertinent, oversimplification: our leaders were right, back in 1958, to elect civilian rather than military control over our Space activities. As a retired Air Force reserve officer and at one time a staunch advocate of placing the Air Force in charge of our more ambitious manned spaceflight programs in the interests of national security, I have not come full circle quickly nor easily. But I find that I am in good company; a brilliant Air Force officer, Lieutenant General Samuel Phillips, who earned his spurs as manager of the Minuteman missile program and who was borrowed later by NASA to head up its Apollo program, gave me two reasons that justify the civilian approach:

"First, in the eyes of the world, symbolically, it was correct for us to go to the Moon as peaceful explorers rather than as soldiers. Second, and this is probably the clincher, if Apollo had been a military program, it's extremely doubtful that Congress would have kept on getting up the money for us to finish the job."

Even riskier than overgeneralizing in a book of this kind is the singling out of individual performers in an endeavor with a cast of some 400,000 artisans at the height of the Apollo program, preceded by untold thousands more whose input before that program began was significant. Two such performers, neither of them household names, loom large in my mind because of their virtual anonymity in ratio to their contributions to the Apollo-Saturn Stack. They are Maxim A. Faget of NASA, who had much to do with the conception and development of a manned spacecraft which would survive a fiery entry into the Earth's atmosphere, and Samuel K. Hoffman, president of the Rocketdyne Division of North American Rockwell for nearly a quarter of a century, who furnished Von Braun with 30 of the rocket engines, from the smallest to the largest, for the Saturn V. These two unsung heroes are among many we shall meet in these pages.

For perspective on Apollo-Saturn and its roots during the years between World War II and 1957, I talked to J. Leland "Lee" Atwood on a Sunday afternoon at his office at North American Rockwell in El Segundo, California. It was the only day of the week on which he could spare enough time, and we were less likely to be interrupted there than at his weekend home near Escondido, where he indulges his zest for sailing and cruising.

V

Waiting for Atwood's arrival, I usurped the receptionist's desk on the deserted third floor of North American Rockwell's general offices, a highrise slab that overlooks Los Angeles International Airport, fetched out my notebook, and refreshed my recollections of Atwood's strenuous life.

From the day that he and the legendary James H. "Dutch" Kindelberger left their jobs as aeronautical engineers for Donald Douglas to strike out on their own with North American Aviation's manufacturing division in 1934, the oddly matched pair had worked as a prodigiously creative team.* Dutch, NAA's new president, personified the persuasive extrovert, a veritable piledriver of a man, whereas Atwood, a mild-mannered perfectionist, remained self-effacing but in no sense subservient in the wings. For 28 years he served Dutch brilliantly as chief engineer and number two man until he rose to chairman of the board, upon Kindelberger's death in 1962.

A fiction writer would have hesitated to concoct the tumultuous wave of successes that awaited Kindelberger and Atwood—or the staggeringly painful blows of fate that lay in store for a proud, sensitive person like Atwood—when the team assumed the direction of North American Aviation's struggling little aircraft operation.

*North American Aviation retained that name until its merger in 1967 with Rockwell-Standard.

They found a single passenger plane on order, unfinished, and no other backlog. But the pair had solid credentials, having contributed at Douglas Aircraft Company to the design and construction of the DC-1 and DC-2 which evolved into that most popular of aerial workhorses, still flying all over the world, the evidently indestructible DC-3 "Gooney Bird."

Dutch went out and landed a million dollar order for the BT-9 military trainer, conceived virtually on a tablecloth during lunch, and built the first model in record time. Later, when the British asked North American to build Curtiss P-40's for the RAF, Dutch and Lee promised instead to design a better mousetrap and get it into production just as soon as they could be bending tin on the P-40. Four *months* from the first drawing they had completed the P-51 Mustang, which eventually dominated the skies over Hitler's Europe. By the end of World War II, a stream of 42,000 trainers, fighters, and medium bombers had rolled off NAA's production lines.

The first question I wanted to ask Atwood was how a pair of dyed-in-the-wool "airplane men" had ended up playing so vital a role in placing Neil Armstrong's crew on the Moon. Space hardware, with its home in the vacuum of the cosmos, is about as far as you can get designwise, one would assume, from airplanes that must breathe and swim in the atmosphere. Put another way, the airplane is subservient to the axiom that "what goes up must come down"; thus the aeronautical engineer has been engrossed in the staying "up." Getting the blasted thing "down," however, has been the gravest concern of the aerospace engineer. The one fights gravity constantly, the other, once his bird is in orbit or beyond, almost negligibly. If the vehicle does not return to Earth, and there is not guarantee that it will, the crew is doomed.

vi

Atwood appeared punctually, attired in blue sport shirt and slacks, loafers, and minus socks, apparently fresh from a day on his powerboat. At 65, he still has a lean, fit look about him (he often climbs the stairs to his office instead of riding the elevator), a shyness and a sadness in his eyes even when he smiles, the high-domed brow reminiscent of a meditative scholar and, despite the accumulated stresses of a lifetime spent under mounting pressures of responsibility, an incongruously young expression. He had aged measurably during the recent Apollo years, yet somehow you could visualize how he looked as a boy—he had a look of trusting optimism.

Leading me to a visitor's lounge adjoining the lobby, the giant aerospace company's demanding boss ran immediately into technological difficulties. The lights wouldn't turn on, perhaps because it was Sunday. Frustrated again after fiddling with an alternate switch panel, he shrugged philosophically and tried the door to his own office. Locked. So was the next office. At a third office we gained entry and settled down for our talk.

I was aware—and so was he, I suspect—of the impotent plight of any captain on the bridge of an empty ship, with all hands gone over the side. The absence of annoyance on Atwood's part was characteristic. He is a humble man. At informal gatherings, he impresses (if that is the right verb) most people as being the least conspicuous individual present. This could never have been said of Dutch Kindelberger, who more than dominated the spotlight—he pulverized it.

"Was it you or Dutch," I asked him, "who had the vision to steer North American toward rockets and Space right after the war?"

"I'd like to answer *me*," he said with an owlish smile, his speech betraying only a vestige of his native Kentucky and of Texas, where he got his schooling. "I wish I could boast that as a boy I stared up at the Moon and vowed that one day I'd help land a man there. The truth is, Dutch and I sort of backed into the Space Age, starting with rocket engines. We expected there'd be a need for propulsion systems that were greatly improved over the

German V-2 rocket, as soon as the military got around to fleshing out our manned bomber fleet with long-range guided and ballistic missiles. In other words, we were no longer referring to Dr. Goddard as a nut."

His disclaimer of foresight seemed overly modest as he went on to recount an early gamble that was to bear fruit, a quarter of a century later, at Tranquillity Base on the Moon. At a time when North American's sales had collapsed to a small fraction of their wartime peak, the bold decision was taken late in 1946 to invest a million dollars of the company's own resources in research and development facilities which, by evolution if not design, were to usher NAA into the aerospace age: construction was begun on a test laboratory for large rocket engines and new propellants in the nearby Santa Susana Mountains, and for the country's most advanced supersonic wind tunnel at the main plant at Los Angeles airport.

"A landmark for us," he continued, "was the contract we won from the Air Force to develop the Navajo—the intercontinental guided missile that never went into production. It was an air-breather, you know, that cruised on jet power, but it was launched by a rocket engine. When it was eventually superseded by the ICBM with a nuclear warhead and the program was canceled, there were (and still are) critics who called Navajo a waste of taxpayer dollars. Actually, there was a triple fallout for us and the government that became more than worth the cost: first, what we learned about rocket engines resulted in propulsion systems for a long family of boosters, from the Redstone, Jupiter, Thor, and Atlas that were built for our missile force and for launching Space vehicles in the 1950's, to the big engines for Apollo-Saturn. Second, we had to get into the inertial guidance business in a big way, with a substantial payoff in later defense programs, and third, we learned a lot about supersonic aerodynamics that was invaluable in our later development of the X-15 and then the XB-70. There was fallout, too, for our Hound Dog missile."

"Most people," I said, "could see the obvious application of the X-15 to Space travel, with the pilot climbing almost out of the atmosphere into a near vacuum, but how about the B-70? It's been called another Pentagon fiasco."

"Both programs," he answered, "were necessary stepping stones toward Space. In qualifying for our contract from NASA to build the Apollo spacecraft, we relied heavily on what we'd learned with the X-15 in a near-Space environment, and from the B-70 in new materials, new welding, and other manufacturing techniques, coping with temperature extremes and the integration of very complex, sophisticated systems. It was a sporty course. Trailblazing. The teams we had on those big jobs under 'goers' like Ralph Ruud, "Stormy" Storms, Dale Myers, and Charlie Feltz—I could name a dozen more—were able to acquire the know-how and the disciplines they were going to need soon to make the transition from building airplanes to building spacecraft."

"Everybody's aware of the dollars those programs cost," I said. "What about the human cost?"

"In lives? Well, there were fatalities, as you know, in all three programs. During the flight test of the X-15. The mid-air collision with the B-70—through no fault of the airplane. Then the Apollo fire on the pad. So I'd say the cost, if you lose people, is always high—even one pilot."

"I was thinking more," I said, "of the human cost from the stresses of meeting deadlines under all kinds of pressure."

His eyes clouded with disturbing memories.

"I can't cite you any fatal heart attacks from that cause, that I could document, but there was a high price in broken health, broken homes, and broken careers from overwork and probably from just plain frustration. You have to expect some pretty heartbreaking setbacks as par for the course."

I knew that many of the lines in his face were engraved there by frustrations of his own. A king-size headache on the B-70 had been persistent leaks of nitrogen gas in the "wet" wings of the exotic bird,* through pinprick holes that were virtually undetectable, which nearly drove NAA's best engineers crazy. The difficulties, eventually licked, struck engineer Atwood in his most vulnerable spot—his professional pride. A worse blow fell when Secretary Robert McNamara canceled further production of the B-70 after Atwood's men had met and exceeded the goals set for the unprecedentedly sophisticated triple-sonic bomber.† The facial lines also bespoke indelible scars left by the death of the three astronauts, Grissom, White, and Chaffee, from asphyxiation in the fire on Pad 34 at Cape Kennedy in 1967, and by the ensuing storm of criticism leveled at North American and its boss as builders of the Apollo Command Module. Allegations of "shoddy workmanship" were sheer torture for a man whose whole career as an engineer had orbited around an undeviating devotion to *quality.*

I suddenly had the feeling that I was sitting face to face with the epitome of the "payer" of the price. Lee Atwood has been as representative as any of the breed of key men in private industry whose companies have combined forces with the government to protect our country with superior weapons, and to give it world leadership in aviation and Space technology. Yet today these men are being castigated in unfriendly quarters as villains of a "military-industrial complex" by critics who never point out that the alternative to a partnership of government and industry in these large-scale programs is to let the government assume the entire burden, as in totalitarian states. Nor does it seem to occur to detractors that dissolution of the government-industry management ap-

*The problem was not "fuel" leakage, but leakage of the inert gas employed to prevent a fire hazard.

†A damaging criticism of the B-70 at the time of cancellation was its potential vulnerability to Soviet surface-to-air missile defenses. Yet in Vietnam today, our aircraft of far lower performance have demonstrated their ability to dodge Soviet-built SAM missiles, literally, even without electronic countermeasures.

proach, which achieved the Apollo miracle on schedule, might well deprive us of the most powerful single "tool" forged since the wheel.

Having already extensively researched the question of North American's role in the fire at the Cape, including one angry book entitled *Murder on Pad 34,* I spared Atwood a revival of the subject (which will be dealt with later), and asked him why North American had been selected over such stiff competition. "There were two principal reasons why we won the Apollo contract," he answered. We had the necessary depth of skilled manpower from previous programs I've mentioned, and we were bidding alone, instead of making a joint proposal like other large competitors. As we learned later, NASA was much concerned with centralizing responsibility rather than spreading or diluting it over several companies, so for that reason alone we started in a strong position. In fact, we might have lost the competition if I'd gone along with Jim McDonnell when he phoned me earlier in the game and proposed that we team up, since McDonnell Aircraft had already built the Mercury capsule.* I consulted with Stormy Storms at our Space Division and he hit the ceiling and said, 'Hell no, my guys would all walk out, we don't need anybody else.' So I turned McDonnell down and fortunately it worked out in our favor."

Curious as to why Atwood, at an age when he could gracefully retire, elected to continue bearing a taxing load, I asked him whether the rewards were still worth the price. After some moments of staring at the carpet he answered:

"In a way, the last ten years have been a prison. There's not much room to maneuver in my situation."

His tone was one of patient resignation. He did not elaborate. To cross the gap, I asked him to tell me something of his work habits.

"I usually get up early and fix my own breakfast, with

*North American won the original competition for a man-in-space capsule from the Air Force, but lost out to McDonnell when NASA was created in 1958 and held a new competition for Mercury.

an eye to the scales—wheat germ and health foods—after a workout on an exercise machine. My business day starts when I hit the office in my T-Bird about eight forty-five. Unless there's a crunch on, I leave around six."

"With bulging briefcase?"

"Not often. I need the evenings at home to put my feet up and recharge the battery."

"I notice you're not smoking."

He gave a quick, shy smile, proud of himself.

"Couldn't seem to get much below four or five cigarettes a day for a long time," he said. "But I finally made it."

As a melancholy afterthought, he added, "Cigarettes killed Jan, I'm convinced."

He was referring to his second wife, who died of lung cancer while still young and very beautiful, a blow which befell NAA's chairman in 1964 at a time of mounting strain in meeting the company's Apollo commitment. Some friends feared that he might go under.

"He looked like a dead man," one said. "It was a godsend that he decided to remarry (in October 1968). He's been like a new person since."

"How much of an inroad on your time comes from entertaining chores?" I asked.

"Practically nil," he said with satisfaction. "Government regulations for defense contractors forbid most of it, so I've got the perfect out—I'd rather spend the time with my family or working."

The policy has been somewhat relaxed, but until recently Atwood set a Spartan example for the other executives by flying the airlines tourist class. He eschews the use of company aircraft for personal convenience, pays his wife's fare and hotel bills even on official business trips abroad, and defrays incidental expenses like taxis and tips out of his own pocket. I asked him why he leaned so far over backward.

"When you're in a position to approve your own expense accounts," he said, "you've *got* to lean over backwards. And the smaller items take too much book keeping to bother with."

He shot me a look that was half serious, half joking.

"Besides, Beirne," he said, "I've always had a feeling I was being overpayed to start with."

On this note of frankness, he rose, apologized and said he was expected at home, and asked if we could continue the interview in my car. His wife had dropped him off, and he was without wheels.

During the short drive to the apartment facing the ocean at Marina Del Rey, where he resides during the workweek, he spoke of the formidable obstacles that had been overcome by the builders of the engines for Apollo-Saturn and of their relative obscurity compared to the better publicized builders of spacecraft.

I assured him of my intention to explore beyond the domains of better known leaders such as Von Braun.

"Do right by Von Braun," he said. "I have enormous respect for him. But go see Sam Hoffman and his team over at Rocketdyne, too."

"I have an appointment with him," I said, "for tomorrow."

THE SEMIFINALS (1950's)

"The longest journey begins with a single step."
Chinese Proverb

Canoga Park, California, home of the Rocketdyne Division of North American Rockwell in the western basin of the San Fernando Valley, bears little geographical resemblance to the sandy dunes at Kittyhawk, North Carolina. Nor is there any monument, other than a complex of olive drab structures for the manufacture of liquid rocket engines, comparable to the white granite shaft that commemorates man's first powered flight by the Wright brothers.

There should be. And for a kindred reason.

At Kittyhawk man first used power to fly successfully in the atmosphere. At Canoga Park he took his final upward step by conceiving and perfecting the awesome power plants needed for manned flight to the Moon. For centuries the stubborn obstacle to controlled flight had been less aeronautical ignorance—men had learned how to soar aloft in balloons and gliders—than the absence of a suitable engine. The window-rattling thunder from rocket test stands in the adjacent Santa Susana hills during the Fifties was the birth cry of Apollo-Saturn.

What manner of man, then, I wondered as I approached his reception desk, would Samuel K. Hoffman turn out to be?

On the precise second of our appointment, Hoffman emerged from his office to greet me.

"I've been looking forward to shaking your hand," I said expansively.

It was a blunder. Sam leveled a steady gaze at me.

"You've shaken it before," he said.

Embarrassed, I followed him into his office, striving to recall some previous encounter.

"Well," I apologized lamely, "how *not* to begin an interview."

"Not at all," he said. "It was some time ago, in a group."

Later I was told that he is adept at needling people whom he likes, so I felt better.

Sam Hoffman is a small man, and, when he removes his glasses, handsome. His eyes, warm and alert, give you the feeling that little escapes him, that he has always been an astute observer, whether the object might be a nozzle injector or a fellow being. Or a production curve. It is fortunate for the Space program that Sam is not a larger man. Otherwise, he would never have been able to squeeze through a small window he broke open in an airliner that crashed—jamming the emergency exits—in 1943 at Bowling Green, Kentucky, in a violent electrical storm. Sam escaped from the burning wreck with a concussion and a broken ankle. There was only one other survivor.

Five years later a second quirk of fate preserved Hoffman's destiny—a phone call out of the blue—quite possibly with a bearing on the future timetable of Apollo-Saturn. This is not to assume that under a different man's aegis Rocketdyne would necessarily have failed to achieve breakthroughs with mammoth rocket engines, but unquestionably the Division was to benefit from a rare continuity of leadership from Sam over a span of twenty years, and from the principal managers he selected.

The long distance call came in 1948 at an unpropitious juncture when Sam had ensconced himself in an Ivory Tower at his alma mater, Penn State, where he had become a professor of aeronautical engineering. The caller, Dr. William Bollay, offered him the post of chief of the Propulsion Section at North American's Aerophysics Laboratory (which was to evolve into Rocketdyne). Content in his academic niche, happy in a full home life with his wife Genevieve, a charming, calm and poised brunette, and in the prospect of ample leisure to devote to his family which grew to include four children, Sam did not need a new and arduous job. He did not jump at Bollay's offer. As an engineer and executive with large responsibilities at Lycoming and other companies, he had been well bloodied in the battle of competitive industry—in fact his move to Penn State had been a deliberate change of pace to a peaceful haven.

"I can't give you a decision on the phone," Sam told Bollay. "But if you want to wire me a ticket, I'd be glad to fly out and talk about it."

There being a limit in cloistered halls to the challenges open to a complex individual like Hoffman—part scholar, part adventurous spirit—the "man of action" predominating in his psyche responded. The Hoffmans moved to California, and to a life far removed from academic detachment.

"Well, what can I tell you for your Saturn-Apollo book?" he asked pleasantly, and a mite defensively.

"For a start," I said, "is the right name for the program 'Saturn-Apollo' or 'Apollo-Saturn'?"

"Depends where you're sitting," he said, smiling. "If you're talking to Von Braun in Huntsville, or to me, Saturn comes first. Over at our Space Division, Bill Bergen would put Apollo first, and so would Bob Gilruth or George Low at Houston. Certainly in the press the

Bergen would put Apollo first, and so would Bob Gilruth or George Low at Houston. Certainly in the press the spacecraft gets the play more than the launch vehicle and the engines."

"I guess things haven't changed much since the old days of the National Air Races," I said. "The airplane got the headlines, never the engines."

"Still the same," he said. "And I imagine the Air Force must feel like the rocket men. Navajo, Redstone, Jupiter, Thor, and Atlas—all the launchers for our missile and Space program right up to the Saturn series—all were powered by Rocketdyne engines developed under contract to the Air Force. I'm sure NASA hasn't forgotten the debt, either. The F-1 engine for Saturn V was a direct outgrowth of Air Force programs."

"Was any one individual primarily responsible for the F-1?" I asked. "In retrospect, it seems like rather a bold concept for a design engineer to think in terms of a single chamber rocket engine of 1.5 *million* pounds of thrust, so many years before any requirement for a lunar mission had been established."

"I can't name any one man," he said after a thoughtful pause.

"You for instance?" I asked.

"Whatever I had to do with it," he said, "was sort of in the nature of all engineers, to keep shooting higher. We started off here by uprating the German V-2 propulsion system, which had only 56,000 pounds of thrust and used liquid oxygen and alcohol for fuel, which we eventually changed to kerosene. When we'd worked our way up to 100,000 pounds thrust, we'd start gunning for 150,000 and then 200,000. So it was almost inevitable that my people, with government encouragement, began studies in the million to million-and-a-half range—initially for the Air Force and later for Von Braun, when he was given responsibility for the Saturn V. Now we're thinking in terms of 20 million pounds. And beyond that."

"During those years," I said, "the Russians were ahead

of us in the weight-lifting business. Were you trying to catch up?"

"That takes a little explaining," he said. "Ironically, the Soviets got ahead of us for the wrong reason, because they were behind us at first, militarily. Compared to our nuclear warheads, theirs were big and heavy, requiring bigger engines than we needed. That was fine for them early in the so-called Space Race, adapting all that existing military hardware for a new purpose. They got the jump. But we leapfrogged past them in our civilian program. Just as soon as we knew we'd have to lift 6.5 million pounds off the pad for Apollo, we had a goal of developing roughly 7.5 million pounds of thrust to do the job. A single engine of that size was out of the ball park, but 1.5 million pounds seemed attainable. You could cluster five of the brutes and come up with the target figure of 7.5 million. Of course, the Soviets could have surpassed that, theoretically, by clustering a larger number of smaller engines, but there's a practical limit somewhere."

Our conversation turned to the upcoming Apollo 10 mission (on April 18, 1969), the last rehearsal for the Apollo 11 landing, and to the prospects of success so far as engine reliability was concerned.

"Before our first manned shot with Apollo 7," he said, "I wouldn't have bet a dollar against your thousand that it was possible for us to go this far without a single engine failure in Space, like we had on an unmanned mission, with the J-2 liquid hydrogen engine on Apollo 6. Now I feel much more confident."

"Incidentally," I said, "that was quite a detective story on the J-2—how your sleuths were able to pinpoint the cause and correct it in time to keep on schedule with the manned missions."

"It wasn't easy," he admitted proudly, "with no witnesses and no hardware we could recover from Space. The J-2 man for you to see about that is Paul Castenholz, across the street."

I assured him that the extraordinary episode would occupy a prominent place in my book. We concluded my visit on the note that, rather than tax the boss with too much detail, I would interview some other gentlemen whom he named, in the areas of their specific responsibility. But I had a final question for him, regarding his management philosophy.

"Would you comment on the view I've heard expressed by some successful executives," I asked, "that shaking up a management team every four or five years is a healthy policy? Rather than for one man to be in the saddle as long as you have?"

"I disagree," he said stoutly. "As long as the team is cutting the mustard, why change a winning combination?"

"Well, there's the question of overidentification," I said, "of friendship getting in the way, so that the boss is inclined to forgive shortcomings. For that reason, one division president has told me that he's never been to dinner in a subordinate's home, nor accepted purely social invitations from his own boss."

"*Real* friendship is appropriate in *any* situation," he said flatly. "Including mine. I may be wrong, but I've always felt that the men who've been with me for years are my friends. And they know that I expect more, not less, of them because they *are* my friends. You don't bother to chew out a manager you don't like and don't respect—it's easier to get rid of him. When you bear down on a friend, it's a compliment, it's a way of letting him know how much you've come to expect of him."

I left feeling that I had learned part of what I came to find out: how, in terms of individuals, Apollo-Saturn (or Saturn-Apollo) had been built. Sam Hoffman had turned out to be another in a succession of clues to the larger truth—that there are no great men, there are only great challenges, which ordinary men at the right time and place become inspired to meet. He had the special kind of loyalty which I was to encounter often again—loyalty first to himself and the deepest potential that had been born within him, and loyalty to his colleagues in a vast endeavor.

An example of the latter was given to me subsequently by William Brennan, production manager, who described a tense meeting attended by Hoffman and some of his key managers at which an irate Air Force general raked Rocketdyne over the coals because of delays in qualification of engines for the Atlas missile. He implied in strong language that there was either incompetence or negligence or both. Sam's hackles began rising as the criticism of his subordinates continued. Finally he stood up.

"I've heard enough, General," he said quietly to his most important customer. "If you can find five other men in the whole country who are half as well qualified to lick the problem, and as highly dedicated, as these five sitting here, for God's sake tell me where they are because I can use them!"

ii

Over the next few days I had meetings with some of the Rocketdyne managers whose lives had revolved around building the kind of engines that eventually boosted Apollo to the Moon. Two courses were now open to me. I could bone up on the facts about highly technical hardware, from the viewpoint of an engineer (which I am not), or distill from what these men told me a layman's feel for what had been significant in the recent history of the propulsion "leg" of the three-legged stool referred to earlier.

A middle course seemed best, calling on just enough technical homework to avoid a performance on a bare stage. For the other two legs of the stool, the gyro and the computer, as their relevance to the Space program grew during the Fifties, and for other roots, I shall attempt the same middle road, with an eye more to the manner of men than to the hardware.

William C. Guy, who has stayed the course all the way as Hoffman's number two man, turned out to be an intensely serious man who seldom smiles. He displayed a phenomenal photographic memory for facts, figures, and details when he explained the genealogy of Rocketdyne's family of engines, even twenty years back. As he talked of thrust-weight ratios, expansion ratios, and other esoteric formulae of his trade, his pencil was busy on a scratchpad. Like every other scientist-engineer I was to meet, who within three minutes resorted to a pencil or a piece of blackboard chalk to "talk" to you, Charlie Guy drew pictures to explain his points. If it were the nozzle he was discussing, he didn't just draw the nozzle, he sketched the whole engine—again a universal trait of the fraternity. Their mental disciplines demand absolute *clarity,* no misunderstandings, extreme simplicity, and the relating of the part to the whole. The nut of what I got from Guy was that his organization, borrowing heavily on the work of Goddard and Von Braun, had been given enough time and elaborate enough facilities to learn how to control an explosion. He didn't put it quite this way, but in laymen's terms, when you set fire to potentially explosive liquids, contain the fire and channel it according to your whim, and keep the whole thing going from thirty seconds to five minutes, then you've learned how to fire a bomb off slowly. To me, that's what it had all been about: the nation can take off its hat to the magicians in tin hats who manned the rocket test stands while the Earth shook from their man-made Vesuviuses.

Surely, in his efforts to inform an ignorant author, Mr. Guy was another archenemy of the nonfact. And so, I discovered, were veteran managers Tom Dixon, formerly boss of engineering; Paul R. Vogt; Joseph P. McNamara; William Brennan, later in overall charge of F-1 and J-2 production; Norman C. Reuel, Brennan's highly articulate and persuasive assistant general manager; and Paul Castenholz, manager for the J-2 liquid hydrogen engine.

When the Apollo program matured in the 1960's, their

experience with earlier engines would bear fruit as follows: a 5-engine cluster of F-1's (RP-1 kerosene-liquid oxygen) would propel the Saturn IC first stage at liftoff, 7.5 million pounds of thrust, built by Boeing; a 5-engine cluster of J-2's (liquid hydrogen-liquid oxygen) would propel the S-II second stage, one million pounds total thrust, built by North American; and a single J-2 would propel the S-IVB third stage, built by Douglas.

All of these men shared a trait of uncommon tenacity. Nothing like a routine approach would have solved the maddening problems which burgeoned as engines evolved from 75,000 pounds thrust in the first Navajo engine to ever higher thrust-weight ratios and twenty times that thrust figure in the mammoth F-1. The most persistent headache in kerosene-oxygen liquid rockets, as distinguished from solid fuel rockets or from the liquid hydrogen-oxygen combination, was combustion instability. The phenomenon, easily confused by the layman with an "explosion," because it can be violent, is actually an uneven or oscillating burn that usually results in immediate burn-out of the thrust chamber.

Rocketdyne's distinctive contribution to the Apollo program, these men explained, in addition to mastering the novel liquid hydrogen propellant for the J-2 engine, was achieving a dynamically stable injector for the F-1, primarily through evolutionary design of an injector "baffle" which protects the combustion process near the injector face.

Satisfactory operation of a rocket engine under normal conditions was not good enough for Rocketdyne's engineers (or NASA), so they went much further. To induce instability to the point of catastrophic failure, they detonated bombs, literally, in the combustion chamber in order to arrive at positive assurance that the engine would stabilize *itself* promptly against any artificial or natural triggering influence, and that it would not fail even after they'd done their damndest to *make* it fail.

A second high hurdle which had to be cleared before

Apollo could succeed was a diabolically cantankerous new fuel for the J-2 engine—liquid hydrogen. Stored at a temperature of minus 423 degrees Fahrenheit (near absolute zero), liquid hydrogen presented seemingly impossible challenges peculiar to such temperature extremes. An example: air coming into contact with a fuel line could freeze into liquid air. Metals and other materials turned brittle. How do you design fuel pumps, turning at thousands of r.p.m., that will withstand the shock of nearly instantaneous starts and stops?

Why bother, then, with liquid hydrogen when it is so much more difficult to handle than kerosene? Because it was foreseen that a higher energy fuel than kerosene would be needed to propel the second and third stages of Saturn V in their acceleration to orbital and earth-departure speeds, respectively, and to save weight. Another plus, it is a "clean" fuel compared to kerosene.

Elsewhere in the armed forces and in industry I have listened to harsh criticism of the technology, termed by some as "archaic," that ultimately produced the propulsion systems for the Saturn V. It has been asserted in certain quarters that the goal should have been a "transportation system" with engine reliability of an order approaching that found in airline operations, that solid fuel rockets would have been much simpler, cheaper, safer, and more timesaving, used perhaps in combination with liquid types, and that interservice rivalries and deplorable duplication of effort and resources had resulted, to the detriment of a unified and more imaginative approach.

On the first point, had there been a less pressing time element in 1953, to close an alleged missile gap and later to carry out President Kennedy's Moon-landing pledge, rocket engine technology for a more economical transportation system might have made sense. Against deadlines, however, the goal had to be to get the immediate job done, with long-range utility secondary. Insistence on reliability, in the context of astronaut survival and mission success, was never for a moment compromised and it was to be magnificently achieved, as of this writing.*

On the second point, liquid versus solid propellants, brisk controversy has not yet subsided. At one extreme, "solid" advocates will question, in private, the integrity and even the sanity of those in the "liquid" camp such as Von Braun and Hoffman, implying that they were merely protecting an enormous vested interest in manpower and machines. "Think of all the people they'd have had to fire," is a typical comment, "if you eliminated all the valves and pumps and plumbing. Relatively, a solid rocket is almost as simple as burning a candle."

Objectively, it would appear that innovation was judiciously sacrificed to urgency. Liquids had been under development for half a century, so they went with the starting lineup which had more experience; it was later to be a keystone of NASA policy to call on current technology for Apollo so far as possible. Solids were a newer technology.

On the third point—rivalry between the Air Force, the Army, and the Navy in missilery, and between the Air Force and civilian agencies before NASA was given the predominant role in Space after 1958—emphatically, there were duplication and parochialism in the development of propulsion systems. As an intermediate range missile, for example, the Army and Von Braun championed the Jupiter, the Air Force, the Thor. Both were produced, a decision which appalled at least one missile expert who was asked by the Department of Defense to list his choices in order of merit. His recommendation?

First choice: Build the better missile.
Second choice: Build the other missile.
Third choice: Build neither missile.
Fourth choice: Build *both.*

*Astronaut survival and mission success are not, of course, synonymous. You can have the first without the second.

With manned space flight in mind, the Air Force planned zealously through the Fifties toward very large boosters right up until the eve of NASA's mandate for Apollo. When the smoke had cleared, however, and whether or not duplication was desirable, NASA had all but one of the booster tools it was to need: the Redstone booster for the Mercury suborbital flight of Commander Alan Shepard, developed for the Army by Von Braun at Huntsville, Alabama, using Rocketdyne's canceled Navajo engine; the Atlas booster for the Mercury orbital series, employing the engines built by Rocketdyne for Convair under Air Force contract; the Titan II booster for Gemini, built by Martin-Marietta for the Air Force with liquid fuel engines built by Aerojet General; and for Apollo, the monster F-1, initiated by the Air Force.* Only the J-2 liquid hydrogen engine bore the label of NASA from the outset; its technology, aside from a novel propellant, rested on previous Rocketdyne experience.

In any event, a strong case can be made for duplication of effort, because it involves competition in conceptual design of hardware. World War II rivalry between the Army Air Forces and the Navy in aircraft and engines, for example, gave this country fighting hardware that was superior, in its field, to the tanks and rifles that were produced under the "arsenal" concept by a single service. Just as no weapons buyer wants to be at the mercy of a sole source, the country may benefit in the end from entirely different approaches to the same goal despite larger outlays. The liquid men gave us the engines we needed in time for Apollo. The solid fuel men have given us a new generation engine for Minuteman and Polaris missiles, and are by no means out of the race for post-Apollo ventures in Space.

iii

No roster of men with the kind of strong shoulders on which the astronauts rode would be complete without

*The contract actually to build the F-1, however, came from NASA.

the name of Bernard "Benny" Schriever, a young Air Force colonel (later a full general) who was placed in charge of our Ballistic Missile program in 1954. Out of the crash program which Schriever headed came all of the boosters which eventually were either converted or further developed for the civilian Space effort; during that period, Schriever was fortunate in commanding the services of diligent researchers, propulsionwise, like Colonel Edward N. Hall of the Air Force's Powerplant Laboratory at Wright-Patterson AFB, whose aggressive approach to any problem at once goaded and encouraged industry during the development of both liquid and solid fuel engines. A hard man to deceive, Hall was of the persuasion that "figures don't lie, but liars can figure." Definitely a born antagonist of the nonfact.

Another fallout from Schriever's Herculean assignment was a new concept of systems management which provided a precedent for the Apollo effort and a training ground for managers of the caliber of Lieutenant General Samuel Phillips; as a young colonel, Phillips had learned how to face up to make-or-break decisions of chilling proportions, managing the Minuteman program, before he was borrowed by NASA for direct supervision of the Apollo program.

For "Benny" Schriever, in conjunction with Rocketdyne during the Fifties, and for Wernher Von Braun during the Sixties, the Holy Grail took the elusive form of Reliability.

How was it, then, I asked Norm Reuel one day in Canoga Park, that there seemed to be such a discrepancy between the incessant and truly prodigious effort that had focused on reliability assurance, at great cost, and some of the blistering letters from the customers which I'd been allowed to see: complaints of quality defects, deviations from specifications, and contamination on engines delivered to be mated with their respective vehicle stages—after the most rigorous inspection by the contractor.

"You have to strike a balance," Reuel said. "If cost and time were no object you could station an inspector behind every worker and attempt to ferret out every defect no matter how minor. Even that wouldn't guarantee perfection—there's no such thing. Significant mistakes must be found and corrected, no question, but that's the negative approach. It isn't just a matter of inspection; the emphasis must be on motivating the man at the bench to do the job right in the first place. Overinspection can result in a tendency by workers to depend on someone else to determine if it's good. Nevertheless, rocket engines are made by people, and people are fallible. We do not live in a world of perfection. A properly defined process, adequate inspections and tests, will prevent "important" mistakes from being incorporated in the delivered product. A theoretically 'perfect' product would never get delivered to the customer at all. Bear in mind also that anything found to vary from perfection may be listed as a 'defect,' even though the part may work as specified—or even better.

"A pages-long enumeration of defects in an inspector's report may look terrible on paper, taken out of context, yet come mostly under the heading of 'cosmetics,' as we call it—like a dealer delivering you a car with minor flaws in the finish; the car still runs just as well.

"The best and final proof of reliability, and a clue to the depth of inspection required, is how the engine performs in a 'hot' firing—and we hot fire every engine to prove its performance and reliability."

"You might also mention divided responsibility," I said. "Things that can happen after the engine is out of your hands."

"No question," he said. "There are many operations involving the engine between shipment and flight that involve other personnel.

"The truth is, and NASA will tell you so, that no contractor escapes these critical letters entirely, and we don't really resent them; it keeps us on our toes, taking immediate steps to comply if we're at fault, which is not always necessarily the case, or to help prevent future occurrences, in any case. A week later we may get a letter from the same customer with a generous pat on the back."

Sure enough, on the wall of Bill Brennan's office I happened to see a framed letter from Von Braun congratulating Rocketdyne for the excellent job it was doing overall. It was dated three weeks following a letter from Von Braun in which he had delivered a sharp rap across the knuckles.

iv

In other parts of the forest during the Fifties, contemporaries of North American Rockwell in industry, government agencies, and universities were also nourishing the roots of the Apollo tree. The emphasis on one company in this book is an unavoidable reflection of North American Rockwell's predominant role in providing propulsion for the Stack and as prime contractor for the Apollo spacecraft itself. It is the only major contractor whose hardware is present in every segment of the Stack, from top to bottom.*

Casting a long shadow, and cut from the same professorial mold as Hoffman, was another man of action named Charles Stark "Doc" Draper, of M.I.T. An aggressive, button-nosed little gamecock of a man, Draper has been obsessed from boyhood by a voracious curiosity. In a sense, he never graduated from college—he continued to take courses for credit for 22 years, even after he had become chief of M.I.T.'s Aeronautics and Astronautics department and founder of its venerated Instrumentation Laboratory. Where most men would be content with

*The eight segments are, in order, the LES (Launch Escape System), the Spacecraft Command Module, the Service Module, the Lunar Module, the IU (Instrument Unit), the S-IVB third stage, the S-II second stage, and the S-IC first stage.

eminence in one field such as physics, Draper thirsted for competence, or better yet mastery, in every even remotely related physical science. Consequently NASA was to be blessed with a one-man technological treasurehouse when the crucial question was posed: can we depend upon having an utterly reliable guidance and navigation system by the time we shoot for the Moon?

Almost any other man would have qualified his answer. The precision required to send men to the Moon safely and back, akin to the difficulty of teeing off a golf ball accurately enough to hit an artillery shell arcing across changing fields of gravity, may have been imagined in the brains of mathematicians, but it had certainly never been demonstrated. Had the Doc hedged, Jim Webb and John Kennedy would have been constrained to throw up their hands on the spot. Why waste even a dollar on engines or spacecraft? If Draper were not putting his money where his mouth was, his supremely confident answer would smack of brashness.

"Hell yes," he told Webb. "If necessary, I'll go along and run the damn thing myself."

It was no idle boast. He had called his shots before, back in 1953, when he startled the skeptics and earned the soubriquet of "Mr. Gyro" by a dramatic practical demonstration of the ultimate use to which a child's spinning toy could be put. Not content with the gyroscope's proven application to automatic pilots for control of aircraft and ships, he was convinced that the same "toy" could serve as the master key to accurate navigation over long distances, without help from radio, landmarks, or stars, when married to an electronic computer and a pendulum device to correct for the Earth's rotation. (Navigational corrections at lunar distance, however, would dictate supplemental inputs from star sightings and ground stations.)

Again, as with the rocket engine, a simple law of motion could be exploited to orient a vehicle in Space: that in the absence of friction, a body in motion will resist any change in direction—in this case the gyro's spin axis. An inertial guidance system, accordingly, once "set" like a watch at a chosen starting point, could tell you where you'd been, where you were, and, through the computer, where you were going and when you'd get there.

To prove it, the Doc had his inertial guidance system, cumbersome by later standards, installed in a B-29 bomber with the announced goal of flying from Bedford, Massachusetts, to Los Angeles without a human hand on the controls nor any course correction from a human navigator. No radio aids. No star shots. After a 12-hour flight, part of it in the grip of a 100-mile-an-hour cross-wind, the B-29 arrived over Los Angeles "as briefed."

When I asked Draper if I was right in the assumption that there had been valuable fallout for Apollo from North American's development of inertial navigation for Minuteman and submarines, or Litton Industries' work in inertial navigation systems for aircraft, he brushed the idea aside impatiently.

"Sure, their systems served their purposes," he conceded, "but if you're talking about *real* precision in Space, our lab didn't need fallout from anybody. The egg hatched here first. Mind you, I'm not talking about production of hardware. You've got to give full credit to people like the AC Electronic Division of General Motors, Kollsman Instruments, and Raytheon—they build the Apollo guidance and navigation equipment under our direction—and to all the computer men."

V

Just what is a "computer," anyway? Since the importance of its role in Space is impossible to exaggerate, a brief course on the computer's origin and nature may be helpful, if not mandatory, here.

The computer originally was and still is a device to help man do arithmetic, and to do it faster than when he counted on his fingers and toes, or lined up sticks and stones to assist him in adding, subtracting, multiplying, and dividing. Such a tool was born several thousand years ago upon the invention of the abacus—a gismo with beads sliding on parallel strings in a wooden frame. Even recently, a skilled operator with an abacus in Japan was able to outperform a modern hand-operated machine.

The earliest successful mechanical calculator (or computer), built in 1672 by a Frenchman named Blaise Pascal, could add and subtract directly, but could not multiply or divide directly. By incorporating an additional set of gears, the Baron Gottfried von Leibnitz uprated Pascal's machine to a prototype for all subsequent calculators, in which all four mathematical functions could be performed directly. Its speed, however, and speed has always been the name of the game, was severely limited by its human operator; man had to be cut out of the loop. The barrier was not hurdled until an English mathematician named Charles Babbage hit upon a method of automating the machine by the use of coded punch cards. Finally, thanks to electronics, in 1946 the fully automatic calculator came of age when the ENIAC computer at the University of Pennsylvania solved a problem in addition in one five-thousandth of a second. Today's computers solve the same problem in one *millionth* of a second.

In the absence of so fantastic a tool, even a Doc Draper would be hard pressed, timewise, in solving the known mathematical formulas for the trajectory of an artillery shell traveling only a few miles. And the U.S. census would be hopelessly obsolete by the time the data had been processed.

Thanks to the phenomenal growth of the electronic computer industry during the Fifties, the builders of Apollo were to find themselves in the happy position of filling their market basket off the shelf, at least for large ground computers, a product in which IBM had long dominated the field. Private industry and government bureaucracies had already created a huge demand. Likewise, on-board computers had been developed for civil and military aviation, missiles and space probes. Also fortuitously, dramatic savings in size and weight, so critical for Apollo, would be forthcoming at just the right time, during the Sixties, through substitution of transistors for the vacuum tube and through improved miniaturization.

Since individual "giants" are difficult to enumerate under the amorphous heading of electronics, the computer industry must take a collective bow for that essential third leg of the Apollo stool. Its engineers well know how unthinkable the Apollo mission would have been without the computer's ability to perform calculations at the speed of light, store data in its memory bank, process and telemeter information back to earth from Space and display it for controllers, engineers, and medical experts, and service a global tracking network.

They, in turn, have not forgotten their debt to the early discoverers of electricity, that mysterious force whose harnessing has become so commonplace that we forget that the most learned men in the world still cannot tell you what it *is.* The known applications of old-fashioned electricity may better deserve the adjective "miraculous" than voyages of men into Space.

vi

If the rocket the electronic computer, and the gyro are the three legs of the Apollo stool, then the seat which they support can be said to be the spacecraft. One overriding consideration governed its conceptual design, as it did the design of the nose cone for the ICBM—heating upon atmospheric reentry.

It was logical that the wax-fastened wings of Icarus would melt when he chanced to fly too close to the sun and plunged to his death in the ocean. Who would believe, then, that a human could fly within two or three feet of the *surface* of the sun? And survive.

That is precisely the magnitude of the challenge that faced the builders of manned spacecraft.

Studying aerodynamic heating during the Fifties, researchers at NACA and other centers calculated (and accurately) that during a typical return of an ICBM from a high trajectory into Space the temperature in the shock wave piling up ahead of the nose cone could reach 12,000 degrees Fahrenheit. That is 2,000 degrees *hotter* than the sun's surface, and ten times the skin temperature that was calculated for the reentry of the hypersonic X-15 rocket plane from near Space. Yet a way would have to be found whereby a man lying on his back only inches from such an inferno would survive.*

At first blush, the "heat barrier" looked insurmountable. Known materials could not undergo such temperature extremes without vaporizing, let alone absorb or dissipate the heat fast enough for the survival of a human riding in the heart of the "furnace." As one scientist put it, it was bad enough to "tee off a Space mission with a man in a can on an ICBM, without asking the astronaut to finish his journey riding inside a blazing meteorite." Velocities for Space flight being what they are, however, there was no alternative but to do just that.

Innovators of fiendish ingenuity were going to be needed. Two such men materialized in the persons of Drs. H. Julian Allen and Alfred J. Eggers of NACA's Ames Laboratory in California, and a third, who followed up their theories—Dr. Maxime A. Faget, joint holder of design patents for the Space capsule which evolved from Mercury to Gemini to Apollo, while he was with NACA and later NASA.

For further insights, I cornered Max Faget (pronounced Fa-zhay) in his lair at NASA's Manned Spacecraft Center, Houston, where he has been serving as Director of Engineering and Development for Apollo

*Capsule designers ultimately reduced reentry heat, in the case of Apollo, to approximately 5,000 degrees—still white hot.

since 1961. As usual, he was working late, so late that our interview spilled over into his homelife.

"I try to keep in shape with a game of squash every night before I go home," he apologized. "I'm overdue now. Do you mind following me in your car and staying for dinner?"

A bit later, trying to keep up with Faget's battered Volkswagen as he drove along back roads to his roomy house southwest of Clear Lake, I was struck anew by the versatility of so many of the "longhairs" I was meeting in the Apollo program. Small (like Hoffman and Draper), brimming with energy, Faget at 48 was at once the erudite scholar, a practical executive who dispensed prompt decisions on hundred-million dollar issues, a dreamer and inventor, and a breadwinner who still managed time for his wife Nancy and their family of three girls and a boy. Dead serious at the core, Max yet somehow conveys a spirit of mischief; it is in his ready smile and his quick, searching glances. I noted also that he has fighting eyes, three-cornered lids, the same kind so often found in pugilists and other aggressive athletes.

Waiting for him in a crisp blue dress half a block from his home was Nanette, age 7. She climbed onto his lap, commandeered the steering wheel, and steered his Bug the final triumphant lap into his driveway—apparently a ritual.

Before we got to the subject at hand—the battle of the great Heat Barrier—my host pointed with pride to his new color TV set, which was working fine.

"I assembled it myself on spare time," he said. "As you know, we're developing a color TV camera to take to the moon. When the experts briefed me yesterday, I was loaded for bear. They were dumbfounded by some of my questions and they're probably still wondering where that little rascal learned so much."

Max gave generous credit to Julian Allen and others at the Ames Laboratory for the breakthrough in hitting upon the right shape for a Space reentry vehicle, in the early Fifties. He told me how Allen deliberately divorced his thinking from past aeronautical experience, which dictated a streamlined shape to reduce friction, hence heating, of objects traveling through the atmosphere at high speeds.

Convair engineers assigned to the Atlas program, calling upon mountains of prior aerodynamic findings, had fed the problem into a digital computer. Not surprisingly, it came up with streamlining as the best answer. On the assumption that Convair might have cut off its computer too soon, Allen pursued his hunch that streamlining could be the worst possible answer. Unlike the goose, which remains in the trap it has walked into because it never occurs to it to back out, Allen reversed the direction of his past thinking. He began reasoning that streamlining gave you two things that you didn't want, for either nose cones or spacecraft.

First, it would allow the reentry body to plunge through the atmosphere faster than a blunt object, say a sphere, when the goal was to apply the brakes; and second, streamlining would send a smooth flow of air past the sides of the object, exposing it to the scorching heat the air had absorbed. Ergo, go the opposite route. Design a body that would precipitate maximum air resistance, or drag, when it streaked back into the "sensible" atmosphere (about 400,000 feet), thus slowing it down, and that would exploit a shock wave's behavior—like a "bow" wave it could both separate and insulate the vehicle from the intolerable heat and also deflect the heat away from the flanks of the beast. In other words: concentrate upon experiments with blunt bodies, perhaps cone-shaped, traveling blunt end first.

Faget, and his confreres under Dr. Robert Gilruth at Langley Field,* followed up the Allen-Eggers blunt-body lead in their conceptual design of the Mercury

*Among them were Robert O. Piland, Kenneth S. Kleinknecht, and Caldwell C. "Cad" Johnson.

capsule, but an almost equally formidable obstacle remained. That blunt end, breaking the trail, as it were, through a mass of stationary air molecules, would need a pretty tough hide—tougher than any heat shield yet invented—if it were not to be incinerated along with the astronaut.

Two approaches seemed possible: a "heat sink" of beryllium or similar metal capable of absorbing the heat; or layers of ablative material like fiberglass and resins, which would char and peel off, but applied thickly enough so that the coating would not be burned through to the spacecraft's skin. Both were incorporated in the original Mercury capsule, but ablation won the day for later spacecraft, with technology furnished by Avco Corporation. Through ablation, you were able to discard the heat instead of merely absorbing it (and worse, retaining it), and you saved weight.

"Did the Russians 'borrow' the concept of a blunt cone from us?" I asked Faget.

"We've never been on their mailing list," he smiled, "but I understand that they started with a sphere, which of course is also a blunt body with certain virtues for ballistic reentry. Apparently they did borrow from us later for their Vostoks. I don't really know."

Once more, men in dogged search for facts, however inimicable to past experience, had kept alive the dream of going to the Moon. It took one man, like Allen, to hit on a radical solution to a problem which threatened to stop the ICBM program cold in the middle Fifties. It took another man, like Faget, to grasp the implications for manned Space flight and act vigorously on them. We will hear more of Max when the story catches up with Apollo itself, but meanwhile here is a further observation, from a colleague of his:

"Max never tries to make the wheel rounder. He's seen too many perfectionists 'improve' a system into serious trouble, when it was already hacking the mustard. Max's kind of guy can make all the difference when engineers are tearing out their hair, meeting deadlines."

vii

A tree whose technological fruits were eventually plucked for Apollo's benefit was planted in 1936 at Pasadena by a genius named Dr. Theodor von Kármán, namely JPL—the Jet Propulsion Laboratory at Cal Tech. Thanks to the advanced facilities and the distinguished staff of JPL over the years from World War II to the birth of Space flight, invaluable experience had been gained in experimental rocketry, radio guidance for missiles, telemetry (the sending back of precise information by radio code on the performance of man or machine even from cosmic distances), and in the creation of the unmanned satellites and Space probes that were destined to scout the unknown before astronauts could safely follow the same trails.

From JPL, under the direction of Dr. William H. Pickering, came America's first Earth satellite, the Explorer I, launched on January 31, 1958, in belated reply to Russia's Sputnik, orbited four months earlier.* After JPL was acquired by NASA in 1958, it was given direction of the Surveyor, Ranger, and Lunar Orbiter series of probes which told Apollo's planners what they had to know about the Moon before men could be sent there—supplying photographs of the terrain for landing sites, radiation and gravitational data, and geological information on the nature of the lunar surface. You couldn't begin to design a manned lunar lander until you knew what it would be landing *on* and how much weight the surface would support. If any.

A lesser spirit than Dr. Pickering's might have succumbed to the bleak discouragement and the barrage of criticism which accompanied a rash of early failures

*Von Braun's team, under Major General John B. Medaris, boosted the Explorer I into orbit.

before the success of Ranger 7; so many missions were "blown" that at the nadir Pickering's job security was on a par with that of a lion tamer who drinks on the job. But he and his team hung in there until unmanned exploration of the Moon, the too easily forgotten side of the coin, was achieved as successfully as was the more glamorous face of the same coin, manned flight.

viii

As the decade of the Fifties ended, no parts of the eventual Stack of Apollo-Saturn were yet visible, not even the engines, nor was there a firm "Apollo" program. But the tools to build the Stack for a manned moonshot, and men with the know-how to use them, were on hand. What was lacking was a catalyst. Moscow supplied it.

It was as if a dozing Uncle Sam had been jabbed by a pitchfork when Sputnik I's haunting beep astonished mankind as it circled the Earth on October 5, 1957. Repercussions of alarm, hurt pride, and incredulity tumbled one upon the other.

President Eisenhower immediately initiated strong measures to spur the Pentagon into bringing order out of the prevailing chaos of divided authority in our Missile and Space programs. Then he won Congressional approval in July 1958 for the creation of the National Aeronautics and Space Administration, to put an end once and for all to the rivalry between the military and civilian agencies in Space matters. Ike set the policy, still in effect, that this country's manned exploration of Space would be a peaceful civilian adventure, in which the Defense Department would participate as a junior partner only to the extent that might be necessary for our national security, and in unmanned programs.*

NASA, under the direction of Administrator T. Keith Glennan through 1960, and starting with a nucleus of

*The Air Force went entirely out of the man-in-space business upon the cancellation in 1969 of MOL (Manned Orbiting Laboratory).

engineers at the former NACA headquarters at Langley Field, Virginia, quickly absorbed Von Braun's organization at the Army's Redstone Arsenal, Huntsville, Alabama, and rushed preparations for the launch from Cape Canaveral of Explorer I, which orbited successfully four months after the first Sputnik.

Mindful that it was but a question of time before the Soviets would essay replacing with humans the dogs which they were lofting into orbit in follow-on Sputniks, NASA began feverish efforts to tailgate if not overtake the competition. Seven astronauts were selected for intensive training. Faget's group redoubled its drive to perfect a capsule in which an American could at least survive a suborbital flight, and new pressures were applied to get the Air Force's failure-prone Atlas booster "man-rated" for true orbital flight in a proposed "Mercury" series. Plans were projected beyond the one-man Mercury vehicle to the two-man "Gemini" and its eventual successor, a three-man "Apollo" lunar vehicle, although no such requirement had yet been formalized.

This, then, was part of the broad backdrop against which President Kennedy was to appear before Congress on May 25, 1961, the other part being the humiliation of the Bay of Pigs, and the need, politically, for the new Administration to seize upon some bold form of atonement.

COMMITMENT AND RESOURCES

"I believe we possess all the resources and talents necessary."

John F. Kennedy

On May 25, 1961, one month after Mr. Khrushchev had joyfully embraced and congratulated Yuri Gagarin at a massive rally in Red Square for having been the first human to oribt the Earth, President Kennedy appeared before Congress with a message on urgent national needs. One of these, he spelled out as follows:

"I believe this nation should commit itself to achieving the goal, before this decade is out, of landing a man on the moon and returning him safely to the Earth. No single Space project in this period will be more impressive to mankind, or more important for the long-range exploration of Space; and none will be so difficult or expensive to accomplish."

The new president's appeal, though dramatic and politically timely, was in no sense impulsive. It had been preceded by much soul-searching and fact-searching. During his recent campaign Kennedy had taken a strong stand by telling the voters, "This is the new age of exploration; Space is our great new frontier."

He had divined that popular, hence Congressional, sentiments were ripe for even so far-out a goal as landing Americans on the Moon, despite some trenchant minority views on the Hill. Senator J. W. Fulbright, for one, speculated that history might eventually judge America by the manner in which it dealt with the unemployed rather than the unexplored. And Kennedy had listened carefully to doubters such as his Chief Scientific Adviser, Dr. Jerome B. Wiesner, who believed that manned Space flight was being oversold at the expense of other Space projects, and who advised that he "not effectively endorse" the troubled Mercury program he had inherited from Eisenhower, hence risk shouldering the blame for its possible failure.

Kennedy was well aware of the delays and difficulties that had beset Mercury's development up until Alan Shepard's brief suborbital flight just three weeks prior to his appearance before Congress. He had been briefed on the uncertainties still to be resolved during the upcoming orbital series that would rely on a booster adapted from the Atlas missile—there had been five failures of record. (MA-1, the first Mercury-Atlas combination, when launched into low rain clouds on July 29, 1960, mysteriously exploded one minute later. Cause? Undetermined. Nine years later, under apparently identical weather conditions, when Apollo 12 was struck by lightning as it penetrated a low ceiling, there was to be hindsight speculation that MA-1, unmanned, might have encountered a similar phenomenon.) Still very much an unknown quantity were the lessons yet to be learned from the follow-on Gemini series of two-man orbital missions, in rendezvous techniques and prolonged exposure to weightlessness, before there could be serious talk about sending three astronauts on the exceedingly more difficult Apollo mission.

On the other hand, Kennedy's confidence in American technical competence had been recently buttressed by flights of North American's X-15 rocket plane setting new world records in speed and altitude, by the recovery of instrumented packages from Space in the Discoverer program, by the success of the communications satellite Echo I, a huge balloon placed in orbit 1,000 miles above the Earth like a man-made moon, and of course by Commander Shepard's 15-minute arc into Space down the Atlantic test range.

In the end, Kennedy displayed the grit and foresight to bypass the Doubting Thomases and place all of his bets on those among the men-of-facts who believed that Apollo could succeed, and in a decade.

Thanks to the optimism and enthusiasm of a Congress which generously grabbed for the check, the stage was now set for the Apollo pageant.

ii

Although 90 percent of the actual "builders" of Apollo-Saturn, under NASA's considered policy, would be forthcoming from private industry instead of government arsenals like Redstone Arsenal at Huntsville, the broad term as I will construe it must include to an ancillary extent the architects in government, the flight controllers and mission planners who enabled the finished product to perform its function, and the users. The astronauts were to contribute a helpful flow of advice for modification and improvement of the product. The place of these nonindustry groups in the context of the Apollo story will receive attention accordingly.

When Jim Webb assumed the reins at NASA under Kennedy, he became heir to a valuable legacy of spadework bequeathed by his predecessor, Dr. Keith Glennan. But the explosive expansion of NASA inherent in Kennedy's mandate placed him on the horns of a dilemma regarding management, albeit a familiar one in any very large-scale technical program, such as the Manhattan Project to develop the A-bomb or the crash Ballistic Missile program of the Fifties.

Webb's first task was to select proven decision makers,

doers, in all of the high positions where technical background had to be of a comparably impeccable order. But which quality should he emphasize in his choice of architects?

Most scientists are oriented toward contemplation rather than the shouldering of heavy responsibility. Engineers are given to concentrating upon their own special fields, with little inclination to look at a large-scale program like Apollo as a whole. Where would he find managers who couldn't be overwhelmed by their learned subordinates, yet who possessed the executive ability of men who run large companies—and who commanded salaries which he could not pay? Was there such an animal, and in sufficient numbers? There was only one way to find out—demand both qualities and see what happened.*

Webb drew on three sources and chose well.

At the apex, he was successful in enlisting several of the desired rare species, the scientist of national reputation in his profession who has proved also that he can get things done. They were Dr. D. Brainerd Holmes, Dr. Hugh L. Dryden (Glennan's deputy), and another holdover from the Glennan regime, Dr. Robert C. Seamans, whose talents subsequently elevated him to the post of Secretary of the Air Force; Holmes brought in Joseph F. Shea as his deputy, a man whose name was to become associated with both achievement and tragedy. Parenthetically, after Holmes fell out with Webb in 1963, another topnotcher from science-in-industry replaced him, Dr. George E. Mueller; key man Lieutenant General Samuel C. Phillips was brought aboard one year later.

Webb's second source was industry. He recruited managers of high technical competence to fill key slots further down the line, and he delegated part of NASA's

*Webb himself was neither a scientist nor an engineer. Kennedy appointed him NASA's administrator for only one reason and gave him only one directive: "Manage it."

management functions to private companies and institutions.*

A third source of talent, and an unknown quantity, was already on tap in NASA's own house—a nucleus of scientist-engineers, mostly from the former NACA at Langley Field, whose prior years had been spent almost entirely in the research laboratory, the wind tunnel, or the test range in the field of high speed aerodynamics as members of the Space Task Group directed by Dr. Robert R. Gilruth. There were men like "Max" Faget, "Bob" Piland, "Cad" Johnson, "Kenny" Kleinknecht, and, in the arts of the flight controller and mission planner, Christopher C. Kraft, Jr., and Eugene F. Kranz. Sharp types, all.

In one of the more unpredictable and happy developments of the Apollo saga, previously "cloistered" individuals such as these found themselves able to rise to the new demands placed on them as managers. Gilruth has run the Manned Spacecraft Center in Houston all the way, with Piland, Faget, Kleinknecht, and Johnson serving ably in major capacities.

NASA's other two largest centers, Marshall Manned Space Flight Center in Huntsville, Alabama, and Kennedy Space Center at Cape Canaveral, Florida, have been continuously under the aegis, respectively, of scientist-engineers Wernher Von Braun and his former subordinate Dr. Kurt H. Debus. Since Huntsville had been a customer for military hardware through the Fifties (when it was called Redstone Arsenal and then ABMA—Army Ballistic Missile Agency), Von Braun and Debus held the advantage over the Langley group of having been exposed to the administration of large procurement programs. Both men, additionally, had had previous management experience at upper levels in Germany's V-2 program.

*General Electric, for example, was given responsibility for "integration" of the Stack—insuring that all participants were playing the same tune. Boeing also served as an integrator for NASA.

Equally impressive to me as the adaptability and versatility these Langley men showed, in their transition to an alien way of life, was their surprising foresight—how closely the eventual Apollo reality came to resembling the picture they had conjured in their imaginations a dozen years earlier.

With this in mind, I went in early May of 1969 to see Dr. Gilruth at the famed gleaming-white complex southeast of Houston from whose innards emanates that familiar TV announcement: "This is Mission Control."

iii

"Your timing is awkward," a NASA public relations man told me. "With Apollo 10 around the corner, Dr. Gilruth and everybody else here at MSC are in a tight bind, but I'll try to squeeze you in."

Luck was with me. It turned out that Dr. Gilruth owned a copy of a book I had written 34 years previously—*I Wanted Wings.* An ardent flying buff, he wanted it autographed. Perhaps because of this fortuitous link with the past, I found MSC's harried director affable and warm, by contrast with reports I'd heard that he tended to be inaccessible, or, when he did grant interviews or press conferences, un-newsworthy. He pondered questions studiously but answered honestly, I thought, and constructively.

"If you could go back ten years," was my first question, "what, if any, changes in the Apollo program would you make?"

"You mean really major changes?"

"Yes, the basic guidelines for building the Stack and flying the mission."

"None," he said, "except the changes we made in the Command Module after the fire on the pad. Our early concepts and assumptions have stood the test well, including all aspects of the Command Module's fundamental design."

He summarized what those assumptions had been

when he and his deep thinkers in the Space Task Group used to indulge in shoptalk about a manned circumlunar flight (minus the landing) prior to their Apollo directive.

First, how many crew members should there be? Until that point was settled, you couldn't determine the size of the spacecraft, which, in turn, would dictate the design of the remainder of the Stack, including the rocket propulsion system.

Obviously, one astronaut would be ideal from the standpoint of reducing weight, complexity, rocket power, and fuel requirements, but equally obviously, he'd need relief on a mission as long as the fourteen days which they correctly visualized as a desirable capability for the spacecraft. Even if some kind of superman were able to catch enough sleep while his bird went on automatic control, there would be no safety margin if he became ill or otherwise incapacitated. Further, Gilruth's coterie reasoned that fully automatic control by ground and onboard systems, with the astronaut "going along for the ride," was the wrong approach, and for the same reasons they had rejected the concept for Mercury and Gemini. They were sure that human judgment, and onboard control of all systems by the astronaut, were indispensable in surmounting unforeseen situations. A computer can never know more than it has been told beforehand. Control from the ground and automation, they concluded, should be in the nature of backup and assistance to the human pilot.

The ideal number they arrived at was a crew of three: a command pilot with overall responsibility, a second pilot chiefly concerned with guidance and navigation, and a third pilot concentrating on the performance of all of the spacecraft's systems—fuel, oxygen, electrical power, communications, and environmental control.

They further visualized that the crew compartment, or Command Module, should sit on top of the Stack so that a rocket-propelled escape tower could yank it to safety if trouble developed during launch. The cabin should be pressurized with its own atmosphere, so as to give the crew a shirt-sleeve environment in which to move around, relax, and exercise. The blunt-body cone, already being proved out on Mercury, should have an offset center of gravity, so that during reentry its attitude could be adjusted by small reaction thrusters, enabling its underbelly to generate a token amount of lift and provide for a partially controlled reentry toward the desired point of splashdown.

They ruled out the concept of a fully maneuverable vehicle, like the winged Dyna-Soar being sought at the time by the Air Force for a reusable Space "shuttle." The latter's weight, limited passenger space, and development time, they believed, posed too many thorny problems for the near term. They agreed on the inevitable need for liquid hydrogen propellants for the upper stages of the launch vehicle rather than the familiar workhorse fuel, kerosene, which would be retained, however, for the first stage; the latter's unprecedented thrust posed a big enough challenge without compounding the problem by a new fuel. If kerosene were chosen for all three stages, however, the Stack could never lift its own weight off the ground. (One gallon of kerosene weighs over thirteen times as much as a gallon of liquid hydrogen.)

Only one major question, that of choosing the best way to actually land men on the Moon, remained unresolved in their minds, and it was not to be resolved until 1962, after all other Apollo guidelines had been firmed up by orders to industry—a story in itself which will be dealt with shortly. Gilruth was to advocate the daring solution finally adopted—the Lunar Orbit Rendezvous (LOR) technique championed by his former associate at Langley Field, John C. Houbolt.

Gifted as he was by foresight and keen insights, it was not difficult to understand why Robert R. Gilruth had been chosen to preside over the procurement of the Apollo spacecraft, the selection and training of the astronauts, and the carrying out of mission planning and flight operations. In this man with the sad, fatherly smile, I had met another trailblazer who didn't just "guess right."

He knew.

iv

If the human resources, the "talents" of which Kennedy had spoken with confidence, were to show themselves equal to the Apollo challenge, what of the technological resources?

A first and immediate concern had to be to get the moonship off the ground, literally. No adequate booster existed in 1961. Neither Jim Webb nor anyone else could predict with certainty that the 7.5 million pounds of thrust he would need before the late Sixties was a realistic goal. The comparatively puny 400,000 pounds of the Atlas represented the pinnacle of booster power to that date. As Webb has said of Kennedy's goal: "It required full and complete success in the development of a booster large enough to do the entire job. . . . Even a small shortfall in booster performance could only spell failure to meet the goal. . . . For a large-scale endeavor, as for the lunar project, a partial success is likely to be a complete failure." (Webb's point is not as obvious as it may seem; had Kennedy elected any goal short of a manned lunar landing, such as a large Earth-orbiting workshop now being undertaken as a follow-on program, compromises could have been made to achieve at least partial success. But there's no way to land a thousand miles short of the Moon.)

The needed hardware resources had been evolving from North American's test stands at Rocketdyne, as has been described earlier, and at Huntsville in the restless brain of Dr. Wernher Von Braun. Until I made my inevitable pilgrimage to the shrine of the patron saint of Space travel (at least in the minds of the public), I had little conception of Von Braun's actual contribution to getting Apollo off the pad.

Had he been the self-promoter that some deemed him, grabbing for headlines and disproportionate credit? Or the selfless servant of a relentless vision?

I admit to a preconception that he belonged in the second category. It seemed implausible that a man driven largely by personal ambition would ever have exposed himself to arrest, imprisonment, and possible torture by Heinrich Himmler's SS and the Gestapo, as Von Braun did in the throes of World War II. The episode is described in a book he co-authored with Frederick I. Ordway, *History of Rocketry and Space Travel:**

> In February, 1944, Von Braun was called to Gestapo headquarters in East Prussia, where Heinrich Himmler tried to coerce him into deserting the army and working for him. Von Braun turned down the proposal and left. A few days later, at 2 A.M., he was arrested by three Gestapo agents. After two weeks in a Stettin prison, he was charged by an SS court. The accusations: He was not really interested in war rockets, but was working on space exploration; he was opposed to the use of V-2's against England; and he was about to escape to Britain in a small plane, taking with him vital rocket secrets. Dornberger went directly to Hitler and said that without Von Braun there would be no V-2. Von Braun was released. Thus Dornberger rescued his brilliant subordinate from a potentially harsher fate than a blasted career.

Equally implausible, it would seem, was the likelihood that any mere opportunist, after switching from German to United States citizenship, could have inspired the trust of his erstwhile mortal enemies, as Von Braun succeeded in doing after the war, first as an adviser to the U.S. Army at El Paso, Texas, and later as a high level director of our own missile and Space programs at Huntsville. Rather, these two considerations alone imply a missionary zeal.

* *History of Rocketry and Space Travel,* by Wernher Von Braun and Frederick I. Ordway III. Copyright © 1969, 1966 by Wernher Von Braun, Frederick I. Ordway III, and Harry H-K Lange. Reprinted by permission of Thomas Y. Crowell Company, Inc., publishers.

Showmanship? Admittedly. A limerick I'd heard attributed to the Sage of Huntsville goes, "Early to bed, early to rise, work like hell and advertise." Von Braun had advertised his theme of Space travel over the years by a prolific flood of articles, speeches, and books.

V

The Russians, like the Americans, had "liberated" a prize catch of German rocket experts in 1945. Accordingly, a witticism made the rounds in Washington, after Russia's disconcerting two "firsts" with Sputnik I and Gagarin, that we had "captured the wrong Germans."

No such sentiments were current in 1969 when I arrived at the Marshall Manned Space Flight Center at Huntsville, to clarify my appraisal of Von Braun's role in making good on Kennedy's blank check for a lunar launch vehicle, or "carrier," as it was referred to at Huntsville. It soon became obvious that the good Herr Doktor was held in something like awe, not only by the "German Colony" of scientist-engineers who had remained with him from the Peenemunde days, but by American general officers and contractor reps assigned to the Center.

My interview started 30 minutes late, for a reason which was given to me afterward at dinner by my German hostess.

"Wernher has a terrible time waking up in the morning," she said. "He reads 'til all hours, oversleeps, and is already way behind a crowded schedule before his first appointment."

Von Braun rubbed tired eyes, visibly drew in a deep breath, and rose from behind his desk to greet yet another visitor, aware that a dozen others were restlessly pacing the outer office behind me. At 57, the striking good looks of his youth had not entirely lost the battle to a lifetime of chronic overwork and crisis or to overweight.

He spotted the bulky, beautifully bound copy I was carrying of the book from which I have quoted. Brightening, he reached for his pen, asking if I had read the volume.

"Not yet, I just got it," I said, "but I've glanced through the index. I was surprised to find only three references to Rocketdyne and none at all to Dyna-Soar. I'm sure you've had a keen interest in a winged Space shuttle vehicle like Dyna-Soar."

His face fell.

"I can hardly believe it," he said. "Of course I've been interested in the Dyna-Soar approach. As for Rocketdyne engines, their importance to us here goes without saying."

Although the book had been published four years earlier, I couldn't resist wondering, not how much my host had helped Ordway in writing the book, but how much of it he'd had time to read.

"I think you'll find," he added, "that the index is misleading, that engines are referred to mostly by their designation, like F-1, rather than by the contractor's name." (This turned out to be true.)

He sat down and autographed the book, to my astonishment, "To a great author from a little one," then smiled, leaned back, and asked what my book subject would be.

"The manner of men who built the Stack," I said.

"You have my blessing," he said, showing quick pleasure.

"But before we get into rocketry," I said, "may I ask you a philosophical question?"

"Certainly, that's the kind I'm usually asking myself," he said.

"Well, the more I've burrowed into this assignment," I said, "the more I've begun to think of the human body as analogous to a spacesuit—an ingenious device to enable each of us to survive on this planet in a hostile environment. Sort of a complete life support system, with our soul and our brain akin to an astronaut inside his spacesuit on the Moon or Mars. Do you ever think of it that way?"

"Definitely," he said. "A good analogy."

"But there's one flaw," I continued. "The astronaut is in excellent communication with Mission Control back on Earth. We, on the other hand, don't seem to enjoy reliable communication with Mission Control somewhere out there in the Universe."

"I disagree categorically," he said at once. "Through all of my life I've been in touch with Mission Control out there, as you put it."

Then he was off. We never did get back to the subject of rocketry until the end of the interview, and then only obliquely.

A serious student of the Old Testament as well as the New, he drew a parallel between the Zionists with their Jehovah and the Disciples with their Christ. With both sects, he reminded me, there had been troubled uncertainty whether or not to share their religion with others, to evangelize, or whether it was their obligation to preserve it for themselves against the encroachment of outsiders. He surmised that it might become manifest in our time that men on Earth were part of a divine plan to carry the tidings of Christianity to other beings on yet undiscovered planets.

As he expounded on related themes, I was drawn under a spell. Whatever Von Braun's rightful niche in rocketry may be, he impressed me as devoutly religious, as a classical scholar and an erudite philosopher. (I consulted afterward with associates of his who heartily concurred.) Fully expecting to find Ye Compleat Engineer, I found it startling to be listening instead to a man of many, and uncommon, parts. Something else shone through, unmistakably—the bright oriflamme of courage. It was an almost tangible radiation from his personality.

Twice during the interview, which was running overtime, a worried secretary rapped on the door, poked her head in, and pointed frowning at her wristwatch. Von Braun nodded and went on talking. He expressed his concern over bringing down the cost of sending payloads into Earth orbit from $500 per pound to $50 by the technology of reusable shuttle vehicles. And he spoke of the more earthly problems of pollution of the ecology and urban blights and how Space fallout might help.

"A 9-year-old boy," he said, "like my son, is not likely to dream of becoming a 'pollution engineer' or an 'urban engineer.' But he can be inspired by Space programs. By the time he's grown and out of graduate school, having aimed all the time at being a Space engineer, it may occur to him that he has acquired the skills needed to attack mundane problems—skills which will be in urgent demand—and forget about Space. People tend to overlook intangible values from the Apollo effort, the stimulating of youth toward scientific and engineering careers in fields besides Space."

When his harried secretary opened the door for the third time, the Sage of Huntsville waved her away with an exasperated gesture.

"I'm sick and tired," he then said to me in a conspiratorial tone, "of taking orders from that woman. If she knocks again, she's fired."

"Rather than let that happen," I said, rising, "I'll take the hint."

On my way out, I stopped at the faithful secretary's desk and asked her how long she'd worked for Dr. Von Braun.

"Oh, years and years," she said.

Her job tenure, I concluded, was in little jeopardy.

vi

In their book, Von Braun and Ordway describe what happened at Huntsville after 4,600 employees were transferred from the Army to NASA and serious work was begun in 1960 on a launch vehicle, or "carrier," for Apollo, with Von Braun carrying over as director.

> President Eisenhower personally dedicated the center, his visit coinciding with Congressional approval of the full $915 million budget requested by NASA for the coming fiscal year.

A large part of that budget was slated for a carrier that had been conceived at Huntsville before NASA came into being. In 1956, the Development Operations Division at Redstone had begun studies on carriers that would go far beyond anything else that was being planned.

As he worked, Von Braun knew that the ICBM-based carriers would not be adequate for manned missions in Earth orbit, let alone manned Lunar and interplanetary flights. . . . Thus, the vehicle and propulsion engineers set out to see how far they could go with the basic elements of Jupiter.

The idea was to cluster a number of Jupiter engines around Redstone and Jupiter propellant tanks to build a large carrier vehicle that would use to the fullest the experience, hardware, and facilities of ABMA and its associated Army and industrial supporters. From this basic idea came the family of Saturns.

On 15 August 1958, the Department of Defense's Advanced Research Project Agency (ARPA) gave its approval for a research and development program whose aim was a carrier powered by eight uprated Jupiter S-3D engines.* These engines, whose thrust would total 1.5 million pounds, would be mounted on a structure consisting of eight Redstone-type 70-inch-diameter tanks clustered around a single 105-inch Jupiter tank. . . . Some changes were made in the upper stages during 1960. . . . In April, all eight engines were ignited for eight seconds, producing 1.3 million pounds of thrust, an American record.

New facilities were built at Huntsville and at Cape Canaveral, and plans were made for shipping complete first stages by barge to New Orleans and then on to Florida. Assembly of the first flight vehicle began in May 1960, while the Rocketdyne Division of North American Aviation, Inc. and Pratt & Whitney Aircraft Division of the United Aircraft Corporation continued to work on engine improvements.

By 1961, Saturn's role as a test vehicle for the Apollo program, which was to take three Americans to the Moon, had been defined. Two editions of the carrier would be built, Block 1 and Block 2. The Block 1 Saturns would have dummy upper stages and would be fired to prove out the basic concept of the vehicle. The first stage of Block 2

*Rocketdyne's forerunner of the H-1 engine.

> vehicles, designated S-1, would carry more propellants and would be powered by upgraded engines, H-1's, which would develop 188,000 pounds of thrust each. The Block 2 Saturns would have stabilizing tail and stub fins, which Block 1's did not; the later carriers would also have live S-4 upper stages ["live" meaning not dummies], an improved instrument unit, and a dummy, or boilerplate, model of the Apollo capsule . . . the 162-foot-long carrier, weighing nearly 1 million pounds, lifted majestically off the ground on 27 October 1961 in a virtually flawless maiden flight . . . .
>
> The rest of the Block 1 vehicles were fired smoothly during 1962 and 1963. The first Block 2 vehicle was launched on 29 January 1964; its second stage propelled a total weight of 37,700 pounds payload into orbit. Dummy Apollo capsules were orbited in May and September by Saturns SA-6 and SA-7, in flights that showed that the spacecraft and its carrier were compatible. The final three Block 2 Saturns orbited Pegasus micrometeoroid-detection satellites, as the Saturn program ended with an unprecedented 100% successful flight-test record.

To depart for a moment from Von Braun's account, what this mainly boils down to was that between 1956 and 1962 Von Braun was juggling three oranges simultaneously—Saturn programs within Saturn programs—in a confusing performance that was to fortuitously succeed in delivering the mighty Saturn V, when Kennedy needed it. I say "confusing" because of the nomenclature of so many versions of Saturns for a plethora of purposes. Two elaborate families of launch vehicles, the S-1 and the larger S-1B, were developed and flown you might say as understudies to the star, the Saturn V, in programs parallel to the latter while still in its birth pains during the decade. As test vehicles, unmanned except for Apollo 7, they were the unglamorous scrubs who never got into the Big Game, but whose contribution was indispensable to the Varsity. You had the overlapping sequence of the Saturn 1, conceived for a general rather than any specific mission, checking out the components of the Saturn 1B, which in turn was needed to prove out the components of the Saturn V, before it could be safely assigned to manned flight. To continue from Von Braun-Ordway:

> Large as they were, Saturns 1 and 1B were only preludes to an even more powerful carrier that was needed to fulfill the goal that had been outlined by President John F. Kennedy . . . .
>
> To carry out the President's declaration, a NASA-Defense Department Executive Committee for Joint Lunar Study, and a Joint Lunar Study Office, were established. And work on the Saturn 5, the carrier that would take Americans to the Moon, was accelerated at the Marshall Space Flight Center.
>
> On 25 January 1962, NASA approved a development program for the carrier which was given the highest priority. Saturn 5 was to have three stages: the S-1C stage, the S-2, and the already familiar S-4B from the Saturn 1B.
>
> The S-1C stage, developed by the staff at the Marshall Center with the support of the Boeing Company, is to be approximately 138 feet long and 33 feet in diameter. Weighing 280,000 pounds empty, it holds some 4.4 million pounds of liquid oxygen and RP-1 kerosene fuel. Each of its five engines, designated F-1, can develop the 1.5 million pounds of thrust produced by the entire first stage of Saturn 1. Saturn 5's first-stage thrust of 7.5 million pounds makes it the most powerful carrier known to be under development anywhere in the world. It has been turned over to Boeing for production assembly at the huge NASA-owned Michoud plant in New Orleans.
>
> The S-2 stage, developed by North American Aviation, Inc., in Downey, California, is 81.5 feet long and 33 feet in diameter. It is powered by five liquid oxygen-liquid hydrogen J-2 engines producing a total of 1 million pounds of thrust. The third stage is the S-1VB, 58.4 feet long, 21.7 feet in diameter, and powered by one 200,000-pound-thrust J-2. Put together, the Saturn 5 stands 364 feet tall;* fully fueled, it would weigh more than 6 million pounds. It is able to send 47 tons of useful payload to the Moon, or to place a 140-ton payload into low orbit around the Earth. By any standards, it is a monument to modern rocket technology.

*"Put together" apparently means "put together with the three modules of the Apollo spacecraft." By itself, the Saturn V carrier was 281 feet high, not 364.

> Saturn 5 is also a monument to the industrial wealth of the United States. Only the very richest of nations can afford carriers on the scale of Saturn; even highly adanced nations have found it advisable to pool theirresources for much more modest space programs.

What were Von Braun and Huntsville's contribution in retrospect? It depends a lot on where you're sitting.

For example, in the foregoing Von Braun describes Saturn V's skyscraper of a first stage as "developed by . . . the Marshall Center with the support of the Boeing Company." Support? When I tried out this credit line on a representative of Boeing, he smiled and shrugged eloquently.

"Don't get me wrong," he said, "but we *built* that first stage, didn't we?"

I got a less parochial reaction from an executive of Rocketdyne, associate contractor to prime contractor Boeing, as we know, for the largest rocket engines ever built. I asked him a leading question.

"Why shouldn't the roles of Rocketdyne and Boeing have been reversed?" I argued. "Aren't those 7.5 million pounds of thrust the guts of that big first stage? Isn't the engine more important than the fuel tank Boeing built to sit on top of your F-1's—like in an automobile?"

"You're oversimplifying," he said. "Boeing's job was to build the *automobile.*"

Which leads me back to Von Braun. When it came down to rocket boosters for Apollo, he was responsible in the same sense for the whole automobile, certainly as an architect. From where I sit, the architect is also a builder, for without him there will be no building. Also, Von Braun and his associates went further than architecture. All were of the "get your hands dirty" school: build enough of the hardware yourself to be able to talk to the contractor, not just as a theorist, but as a hard-nosed equal.

Thanks to the Sage of Huntsville, Kennedy was to be able to cash his largest post-dated check on time.

vii

In rapid succession in 1961 and 1962, NASA's source selection triumvirate of Webb, Dryden, and Seamans awarded the major contracts that could transform aspirations and uncertainties into hardware reality. They believed that roughly 80 percent of the requirements could be defined confidently, with the remaining 20 percent in the uncertain category. Propulsion and navigation loomed large among the uncertainties. Since these might be expected to become bottlenecks when the Apollo program matured (neither were), they received earliest attention.

Doc Draper's M.I.T. lab was given the go-ahead on G & N (Guidance and Navigation). Prime contracts were awarded to Boeing, North American, and Douglas, respectively, for the three stages of the Saturn V. Propulsionwise, each of these companies had a head start thanks to study and development contracts previously awarded to Rocketdyne for the F-1 in 1958 and the J-2 in 1959.

IBM was chosen to build the Instrument Unit, or electronic brain that sits atop Saturn's three stages, with North American building the structure that housed it. The most sought-after plum, for prestige and for its multi-billion dollar price tag—the Apollo spacecraft itself, comprising the Command and Service modules—went in December 1961 to Harrison A. "Stormy" Storms' Space Division of North American at Downey. It was a surprise choice. Only three months previously North American had won the competition for the S-II second stage of Saturn—a major commitment. Industry observers did not expect the same contractor to be "loaded up" with an additional major responsibility, especially since McDonnell Aircraft Company, having already acquired the most experience in building Space capsules (for Mercury), and three giant aerospace teams were in contention: Martin, GE, and General Dynamics' Convair Division. (North American was not "teamed.")

Crucial contracts for subsystems of Apollo's Command Module were entrusted by Draper to GM's AC Electronics Division, Kollsman Instruments and to Raytheon, for G & N hardware; and by North American to Lockheed for the LES (Launch Escape System), to Northrop's Ventura Division for the parachute Recovery System, to Avco for the heatshield's ablative materials, to Garrett for the ECS (Environmental Control System), and to Collins Radio and Honeywell for communications and control subsystems, respectively. As subcontractor for the critical Service Module engine that would have to restart on the dark side of the Moon with 100 percent reliability if the astronauts were to get home, Aerojet was chosen by North American.

It was decided to continue with the Cape as the best site for the Moonport, with Dr. Debus in charge of enhancing existing Mercury and Gemini launch facilities on a tremendous scale for Apollo, to build from scratch Dr. Gilruth's Manned Spacecraft Center at Houston, and to press on with a truly global network of radio and radar tracking and range stations.

The face of the Stack-to-be was now recognizable except for a missing front tooth—an empty cylinder 28 feet high near the top of the Stack, between the Service Module and the Saturn booster. There was as yet no consensus on what to put into the vacant "can." If you were going to shoot for a lunar landing after refueling in Earth orbit by EOR (Earth Orbital Rendezvous), you would fill that empty module with one kind of equipment. If you were to opt for LOR (Lunar Orbit Rendezvous), you would use the can as a garage for a "bug," not yet designed, which would land alone on the Moon.

Direct ascent to the lunar surface and return relying upon only one vehicle was beyond the capability of Apollo-Saturn, unless liquid hydrogen fuels were used throughout, which would involve too much dependence on new technology, and unless the crew were reduced from three to two astronauts to save weight in supporting systems. A super-rocket like the already discarded Nova would have been needed.

The vacancy in the Stack was not to be filled until late in 1962. Therein lies the tale of LOR's stubborn protagonist, John C. Houbolt, and it was well that he was stubborn, as we shall see.

EARLY UNCERTAINTIES (1961-1962)

"We had to do all of the job, including all of that portion in the uncertain zone, or accept defeat."

James E. Webb, NASA Administrator

In any good mystery drama, the author must work back from a credible ending. Apollo's authors in 1961 were in fair shape. Thanks to the lessons learned with ICBM's and early Mercury flights, they had confidence in their ending—a safe reentry and landing on the Earth. Not so with their Act II climax, the lunar landing, nor Act I, which would have to be tailored to the intricacies of Act II. Neither were to be easily plotted.

To switch metaphors, much as Julian Allen with his blunt-body concept had been heaven-sent in the Battle of the Reentry Heat Barrier in the Fifties, a certain Dr. John C. Houbolt at Langley Field was to arrive like the Union cavalry in the nick of time to point the way in 1961-1962 to victory in the Battle of the Manned Lunar Landing technique, or "Fellows, How Do We Grab Aholt of this Critter?"

"Battle" is the right word for the spirited hassling over the choice among five methods of landing men on the

Moon and returning them home intact. And the problem was in truth an ornery critter.

Looking back, it is not easy to perceive whether or not NASA, having contracted for the Saturn V booster and for the Command and Service Modules in 1961, prior to a lunar landing decision not arrived at until the middle of 1962, was being brashly premature. One wonders whether its leaders had painted themselves into a corner, had begun fabrication of a boat in the cellar without measuring the basement door, or both, and had therefore rendered Houbolt's LOR solution not only providential, once thoroughly analyzed, but almost inevitable.

Most experts I consulted now believe that the United States might still today be far from reaching the Apollo goal if any of LOR's alternatives had been chosen. In fairness to all parties, it cannot be overemphasized that no course was open that did not saddle the decision makers with imponderable risks. The execution of the Apollo 11 landing in 1969 strikes me as no more dramatic than the projection in the fallible vision of human minds of the LOR plan to carry it out, seven years beforehand.

Surely the final choice agreed upon reflects really remarkable professional confidence and just plain grit, despite the sanguine talk at the time that the risks were "within the state of the art." The risks were also inescapable and only partially predictable. Equally inescapable would be the agony, for the conscientious decision makers, if the method chosen were to prove fatal, when an alternative, in the light of subsequent events, could have saved lives, and the mission. As meeting followed high level meeting at NASA in 1961, the choices narrowed down.

Direct ascent to the Moon, at one time championed by North American, Huntsville engineers, and Faget at Langley (but not to the exclusion of Earth Orbital Rendezvous), began to lose favor for several reasons. It would require twice as powerful a booster as the Saturn

V which had already been contracted for, since the Stack would weigh 12 to 13 million pounds at liftoff. Since the entire spacecraft would be a round-trip vehicle, hence heavy, it might break through the lunar crust, or sink deeply into lunar dust (the exact nature of the lunar surface had not yet been determined by unmanned exploration). The Stack might land on an uneven surface, canted like a Leaning Tower of Pisa, tumble over, and maroon the crew for all time. The pilots, lying on their backs atop a bird over 200 feet tall, would find themselves in the awkward predicament, visually, of airline pilots trying to land a jet vertically, tail first. Finally, a "brute force" launch vehicle like Nova might entail an extra decade and perhaps billions more dollars to develop than an alternate mode, calling for one or more Saturn V's.

Added together, these factors appeared to outweigh the strongest argument in favor of Direct Ascent, namely simplicity: no dependence on untried, exotic techniques of rendezvous would be necessary. Early advocates, whose stand had been influenced in part by "enlightened self-interest," proved able to rise above parochialism. The appeal to Von Braun, for example, of a super-rocket which would further highlight the role of Huntsville and pave the way for eventual interplanetary voyages, is understandable. And so is the attraction to a prime contractor like North American of that approach which would bring the largest demand for its brand of hardware without the involvement of additional major contractors; the lunar bug, if it were built, shaped up as a billion-dollar-plus prize, which North American could not expect to receive. It already had its hands full at the Space Division in Downey. But either Direct Ascent or Earth Orbital Rendezvous (EOR) would sharply increase the demands for its engines at Rocketdyne.

A second choice, Lunar Surface Rendezvous, began with at least two strikes against it which no one, in the final analysis, was able to stomach. The method called for landing a cache of fuel and supplies on the Moon with an unmanned rocket to await the arrival of a manned spaceship. The advantage? Rockets much smaller than Nova could be used. Strike one against it was that if the crucial payload of the unmanned rocket were damaged in the lunar landing, there would be no return ticket for the astronauts. Strike two, the astronauts might very likely come down too far off target to reach their supply base and refuel (Apollo 11 was to overshoot four miles).

A third choice closely related to the fourth choice, both of which received prolonged consideration from Von Braun, North American, and other influential arbiters, embraced the technique of Earth Orbital Rendezvous (EOR). In one variation, major segments of the Apollo vehicle would be assembled in Earth orbit, thus drastically reducing weight at launch—instead of one huge vehicle, several building blocks would be lifted. Flaw? The precise timing implicit in so many coordinated launches and rendezvous hookups was stringent enough to subject any veteran launch controller like Rocco A. Petrone at the Cape, or Chris Kraft at Mission Control in Houston, to the probability of having their hair turn white overnight. Further, you'd still wind up with the question marks of setting down on the Moon a vehicle almost as ponderous as Nova.

The other variation of the EOR technique, in which an unmanned "tanker" rocket would be placed in near-Earth orbit, followed by a manned Apollo which would rendezvous and replenish its fuel before setting course for the Moon, likewise posed a formidable challenge: subzero liquid fuels are tricky enough when pumped and stored on Earth, with elaborate ground support equipment. Who, then, would want to volunteer to transfer them between two weightless vehicles out in Space? Weightlessness, of course, would have some advantages, like reducing power requirements, but is hardly conducive to human dexterity. And the scheme demanded the costly launch of two Saturn V boosters for each mission. Wanted, to save both money and time, was the ideal stratagem whereby a single Saturn booster could do the whole job.

ii

Ignored both in Moscow and in the United States, a self-taught Russian mechanic named Yuri V. Kondratyuk had unknowingly scored a bulls-eye with a premise he propounded in 1916:

"The entire spacecraft need not land."

He described correctly how a small landing craft could leave the mother ship in lunar orbit, descend to the surface, and rejoin the larger craft, expanding on his theory later in 1929 in a book published at his own expense, called *The Conquest of Interplanetary Space.* He died in obscurity in 1942, a generation away from seeing his foresight vindicated, but at least two Space scientists, Oberth and an Englishman named Harry E. Ross, gave consideration in the intervening years to Kondratyuk's concept.

As has so often happened with creative minds, John C. Houbolt arrived independently at the same premise as had the obscure Russian over half a century before him. Houbolt—relatively obscure as an engineer in the hierarchy of researchers under Gilruth at Langley Field—had branched out into the study of orbital rendezvous techniques in addition to his regular work in aerodynamic research. He had been given no directive to upset the theories most in vogue in 1960—EOR and Direct Ascent—but his growing familiarity with orbital rendezvous complexities stimulated his thinking until the lightning struck, as it had with Kondratyuk: why land the whole bundle on the Moon? Did Columbus have to drive his vessel up onto the beach in order to set a man ashore in the New World, when a longboat was ample for the purpose?

Scribbling rough but workably accurate estimations of

weight factors on an envelope, Houbolt sat at his desk transfixed when the figures yielded two electrifying possibilities: weight at launch could be cut approximately *in half.* Ergo, a single Saturn V booster would suffice for each mission. His mind raced on to the corollaries of these two assumptions. Nearly all came out plusses. Here, he vowed to himself, was a challenge he could not duck, no matter what the personal price might be. He saw in LOR not only the best approach, but perhaps the only one with a chance of beating the 1970 deadline, and at a saving to the taxpayers.

A Midwesterner with steely gray hair and eyes colored to match, Houbolt fell on the loose ball, which he named LOR, in his own end zone and tried to run it out. Tacklers converged from all directions, among them Von Braun, Faget, and Kennedy's powerful scientific adviser, Dr. Jerome B. Wiesner. "It won't work," or "your figures lie," or "that's no good," were some of the comments at high level NASA conferences in 1961, when Houbolt concluded his briefings on other lunar landing options with his LOR pitch, which at least on one occasion he was pointedly asked to omit.

Houbolt refused to be tackled for a safety. He fought his way out to the equivalent of the twenty yard line by means of an in-depth study on LOR, prepared with the help of colleagues on the Space rendezvous committee which he now chaired, and circulating it in the right places around NASA. One of these was the desk of Dr. Gilruth, who, impressed by the hard, cold figures and logic with which Houbolt had presented (he would have preferred "proven") his case, became an articulate convert. Faget and others in the Gilruth group joined the club for LOR, but executives at the highest levels in NASA still were not getting the word. To listen to Wiesner at the White House or Webb at NASA headquarters from 1961 until the middle of 1962, the party line was still either EOR or Direct Ascent.

Convinced, as a scientist-engineer, of the validity of his facts, and bitterly frustrated by what he deemed to be the closed-mind approach of the numerous panels deliberating on the issue, Houbolt resorted in November 1961 to a drastic action reminiscent of airman Billy Mitchell's revolt against the Brass 35 years before. But whereas Mitchell went over the heads of the War Department in a direct appeal to the public for the recognition of air power, Houbolt at least kept the fight in the NASA family by writing a gloves-off letter (out of channels, however) to Webb's associate administrator, Dr. Robert C. Seamans, Jr.

Describing himself as a "voice crying in the wilderness," and as "appalled at the thinking of individuals and committees," he implored Seamans to extend the most urgent consideration to his thoroughly documented LOR proposal. Here the script might have had a sad and familiar ending for Houbolt, with the firebrand being drummed out of his regiment, had Seamans been more the bureaucrat, more complacent, or less open-minded. Fortunately, he was of the caliber of man willing to listen to a minority report from a conscientious source, however unorthodox its content or manner of arrival.

Seamans was sufficiently impressed, and disturbed, by Houbolt's outburst on paper, to buck it on to his director of Manned Space Flight, Dr. Brainerd Holmes—not perfunctorily, but with a request for serious evaluation. Holmes, likewise by no means sold on any single proposal to date, promptly dispatched a knowledgeable emissary to Langley Field: Joseph F. Shea, his newly appointed deputy. Shea arrived right after New Year's 1962.

Later placed in charge of the Apollo spacecraft office at Houston, Shea was admirably equipped at the early age of 36, by previous experience with Bell Telephone and in the Titan missile program, to stop the LOR ball carriers cold at the line of technological scrimmage, had they fumbled or handed off into a busted play. Instead, by the time Shea left, Houbolt was in the end zone for a touchdown, cheered on by supporters who now numbered nearly everyone in NASA except Von Braun.

iii

Back in Washington, Joe Shea's favorable report on LOR to Holmes and Seamans was followed by several months of computer analysis on a national scale; industry, universities, and laboratories were all called upon for a microscopic examination of every aspect of LOR. By the end of June 1962, mountains of data from the computers had carried the day for the Bug approach, so far as Dr. Gilruth was concerned. He presented his conclusions on July 5 at Huntsville to Von Braun and company at a showdown briefing for his old antagonist, if that is not too strong a word for the rivalry which has always existed between Von Braun's rocket-oriented camp at Huntsville and Gilruth's pilot-and-aviation-oriented camp at Houston.

Norman J. Ryker, a North American engineer-executive who had been assigned temporarily to NASA during the buildup to this meeting, recalls vividly what to him was an historic moment of suspense:

"After Gilruth's team was through with its final summation of the case for LOR, all eyes were on Von Braun, who, in the absence of some kind of rebuttal, would be agreeing by default; in other words, there being no objection, 'motion carried,' for a united NASA position. Seconds passed. Von Braun remained silent, and so did his good right arm, Eberhard Rees. It meant Von Braun was finally on Houbolt's side."

Houbolt has expressed grateful admiration for Von Braun's integrity in swallowing his pride, once convinced, and backing LOR thenceforth to the hilt. That support was urgently and immediately needed against renewed opposition from Dr. Wiesner, who demanded still further review of what was now NASA's official position.

As the President's own scientific oracle, wielding great

influence without commensurate or direct responsibility for Apollo's success, Wiesner had for some time been a thorn in NASA's side. At this particularly trying juncture, early in July 1962, when North American had found itself severely handicapped in its progress on the Command and Service Modules pending a final decision on the spacecraft's configuration, the last thing NASA wanted was another chef stirring his ladle in the broth.

And Wiesner was lacking neither in telling arguments nor in distinguished company. Many others in the more purely scientific community stressed with Wiesner the importance, for example, of including a lunar geologist in the landing party, if the Moon's exploration was to be worth the candle. But how many scientists could be entrusted with copiloting a Bug? Ergo, choose a method that could accommodate a scientist as a passenger. There were other plausible objections. With no experience to be gained, under LOR, in Earth-orbital refueling or vehicle assembly, nor Jumbo booster technology pertinent to follow-on interplanetary missions, the benefits from Apollo would be minimized.

Wiesner's most obvious objection to LOR, and for that matter something of grave concern to all parties including Houbolt, was the indisputable and awesome risk of asking a crew to execute intricate rendezvous maneuvers and orbital changes more than a quarter of a million miles away from Earth—that and betting your whole stack of chips that the single engine on the Service Module would restart three times while the astronauts were traversing the dark side of the Moon, hence blacked out from radio communications with Mission Control, for periods of over half an hour.* Direct Ascent, of course, would eliminate rendezvous in Space entirely, while EOR would provide for recovery of the spacecraft in emergency from rendezvous maneuvers carried out only 100 miles away in Earth orbit, with continuous

*First, to establish an elliptical lunar orbit, next a circular orbit, and finally, to depart the lunar orbit for Earth.

communications between astronauts and experts on the ground throughout the entire mission.

iv

Sitting and shifting about upon a hot seat that continued to grow hotter, as the Wiesner-NASA rhubarb dragged on into the last half of 1962, was, of course, the administrator of NASA, a harrassed Jim Webb. Quite properly, as a nonscientist, Webb customarily remained aloof from technical decisions while the experts, hopefully, were debating their way to some acceptable agreement. But time had all but run out. Even after Von Braun had joined Gilruth's LOR forces, Webb was still treading water with public statements that kept alive the options of EOR and Direct Ascent, presumably because of the prestigious dissent of those in the Wiesner bloc.

Let the reader imagine himself in Webb's predicament. Just how eager would the boldest among us be to step in and bell this particular cat? Who would want to assume the responsibility, if LOR were to be adopted, of cutting off Apollo's crew at a climax of the mission from any possibility of help from a home base where literally thousands of technical specialists would be standing by to furnish information or advice that spelled the difference between life and death? On top of that, how many nonscientists would relish such a palm-sweating gamble in defiance of the man chosen by a president specifically to counsel him on so momentous a commitment?

Or, at the summit, who would envy John Kennedy, during his visit to Huntsville in September 1962, when at that late date Wiesner engaged Von Braun in an abrasive debate on LOR in front of the press and the British minister of defense, Lord Thorneycroft—this two months after NASA's recommendation, as it was now being expounded by Von Braun, had been made clear? shortly thereafter Kennedy broke the impasse.

"You're running NASA," he is reported to have told Webb. "*You* make the decision."

As Harry Truman's desk motto so aptly put it: "The buck stops here"—"here" being in Webb's lap, not Wiesner's. And as in the case of Truman's top scientific adviser, Dr. Vannevar Bush, whom Truman inherited from Roosevelt's circle of mentors in World War II, there was the helpful precedent that even the most illustrious expert may not always be right. The eminent Dr. Bush had forecast shortly after the war, and presumably lived to rue it, that development of an intercontinental missile was too far in the future to merit serious consideration—this less than a decade before strategic ICBM's were decided upon as America's first line of defense, and which the Soviets were to deploy first.

Aware of Dr. Bush's earlier fallibility, Webb may have found it that much less painful to arrive at his decision, Wiesner or no Wiesner. But the predominant factor had to be his confidence in the combined judgment of erstwhile antagonists Gilruth and Von Braun, the two builders who would bear the firsthand brunt of the Apollo-Saturn burden, and of Dryden, Seamans, and Holmes.

He bit the bullet in a letter sent in October 1962 to Wiesner at the White House setting forth NASA's final recommendation in favor of LOR. Not until a month later, however, was there a public announcement that the die had been cast, and that Grumman Aircraft Corporation, Bethpage, Long Island, had been selected to design and build the Lunar Module, or LM, as the Bug was eventually designated.

I have said that Webb bit the bullet. Perhaps it is only right to add that the man who bit it before him, even harder than Von Braun, was Dr. Robert R. Gilruth. As the individual most directly responsible for building the Apollo spacecraft, training the astronauts, and planning the missions, with the incorporation of the novel Bug, Gilruth was the linchpin. Whereas Von Braun could consider himself personally off the hook as soon as his three boosters fell down into the Atlantic Ocean, Gilruth's ordeal would encompass the entire mission from countdown to splashdown in the Pacific.

More than any other man, Bob Gilruth was situated closest to the white hot flame of one of history's most difficult decisions. Let it not be forgotten.

V

If the period of the Great Uncertainty preceding the LOR decision could be said to have bedeviled one manager in industry more cruelly than most others, it must have been Harrison A. "Stormy" Storms, the peppery 46-year-old president of North American's Space and Information Systems Division at Downey, California (later designated and hereinafter referred to as the Space Division).

A *Saturday Evening Post* article about Storms,* published in May 1962, six months after Storms' engineers had won the prime contract for the Apollo spacecraft, did not even mention LOR as a possibility. Yet Storms was supposed to get cracking meanwhile as the driver of a 32-mule train that was following different trails to the lunar landing simultaneously.

If the assignment looked impossible, then Stormy was a logical candidate. He was accustomed to giving the impossible short shrift with managerial karate chops and had proved it repeatedly as a mule driver on the X-15, B-70, and other pioneering programs, when fellow design engineers were close to despair.

Knowing this firsthand from previous experimental programs, Bob Gilruth and his "Show Me" deputy, Walter C. Williams, were favorably disposed toward Stormy and his whole engineering brigade when the final selection of a prime contractor for the Apollo moonship took place in December 1961 in the ballroom atop the Chamberlin Hotel, Old Point Comfort, Virginia. (For the losers, the matter at hand would take all the comfort out of Old Point.)

NASA's evaluators, after subjecting all proposals to a

*"Quarterback for the Moon Race," by Art Seidenbaum.

minute X ray, were in the final reckoning most concerned with the human element. They were buying people—men like Atwood, who headed North American's delegation in person, Storms and his teammates, and North American's depth of aerospace talent. Or comparable people like those at Martin-Marietta, which, as was disclosed in congressional hearings six years later, was first choice initially until Webb, Gilruth, Dryden, and Seamens decided to give the palm to North American by a narrow edge.

Atwood's excellent proposal, tailored to Max Faget's conceptual design of the Command and Service Modules, was in part the fruit of a talent hunt prior to the showdown. Stormy had hired top personnel like his deputy, John W. Paup, a kindred human buzz saw.* He had fleshed out his Space Division with skilled manpower even in some cases before a job was ready for them, besides "borrowing" key experts from other North American divisions. With him from first to last were stalwarts such as Dale Myers, Norm Ryker, and Charlie Feltz. The Apollo proposal that resulted did not try to tell NASA its architectural business by "improving" on Faget's approach. It concentrated on implementing it.

vi

The prize in hand, Stormy immediately found himself inundated by those question marks which plague every new development program, and which in the end, after the fire on Pad 34 six years later, were to be instrumental in costing him his job. They have a direct bearing, often overlooked later, on a tragic denouement.

The first and foremost consideration in building a Space-traveling "home" for three astronauts was safety. The second was weight. For every pound of payload in a 100,00-pound spacecraft, 75 pounds of thrust would be needed at launch; if you added another pound, you were

*Paup suffered a fatal heart attack before Apollo was completed.

touching off a nightmarish chain reaction of more fuel, hence more weight, hence still more fuel. But light weight and safety are incompatible. Safety devices and backup systems mean more pounds. You can't have it both ways. So Stormy was plunged from the outset into tradeoffs—compromises between the two.

Pending a decision on LOR, Stormy could not freeze the design of the conical Command Module. With either Direct Ascent or EOR, no provision need be made for astronauts to transfer to another capsule in Space. But if LOR were chosen, two men would have to crawl through a tunnel in the apex of the cone in order to transfer to the Bug. Fortunately, Stormy's contingency planning was flexible enough to accommodate revisions for a hookup to the Bug when the requirement tardily arose.

Two early question marks which were not to be clarified until after three astronauts met their deaths make Stormy's judgment look good in retrospect. He originally favored a quick-release hatch, opening outward especially suited to prompt escape in case of emergencies on the ground, and, to reduce fire hazard, a two-gas atmosphere in the cabin, part oxygen, part nitrogen. But for plausible reasons which will be elaborated upon later, NASA ruled in favor of a manually-operated inward-opening hatch and a 100 percent oxygen atmosphere, a combination that, as we know, was to prove fatal in unforeseen circumstances. Suffice it here to say that Stormy's approach accented safety at sea level, whereas NASA's emphasized safety out in Space, where inadvertent opening of a quick-release hatch could spell catastrophe, and where fire could be extinguished by immediate cabin depressurization (if nitrogen were present the crew would suffer the "bends"). In any event, the fire hazard lurking in a 100 percent oxygen atmosphere under high cabin pressure, though recognized, was jointly underestimated by NASA, by the contractor, and by those with most at stake, the astronauts themselves. This they were to freely testify, after the fire.

Thermal protection was another design uncertainty. Storms could not project the environmental control system of Apollo until he knew whether the entire spacecraft would land on the Moon, in which case it would be exposed to long-term alternate extremes of cold or heat. If only a Bug were to be subjected to the harsh day-night cycle of temperatures on the lunar surface, his design and manufacturing headaches would be that much simplified at an appreciable saving of weight.

"Weight."

Like the cry of a bird, the word became an incessant refrain for Stormy and the builders of Apollo—a sort of new and final Great Barrier Reef. If Von Braun added a hundred pounds to Saturn's first stage, he'd be rid of it about 2½ minutes after liftoff, at stage separation. But if Space Division added a hundred pounds to its modules, they must be toted around in Space for eight days.

Stormy's battle with the weight barrier was by no means confined to the spacecraft at Downey. His Space Division was responsible also for the S-II second stage of the Saturn booster, being developed under Bill Parker (later succeeded by Robert E. Greer, a retired Air Force general, before the S-II operation was moved to nearby Seal Beach, California). Here again, the prime bugaboo was weight, the saving of which dictated resorting to a pair of extreme measures—the thinnest possible aluminum skin for the 81.5-foot-tall bird, and the adoption of liquid hydrogen as the propellant, sharing a common bulkhead with the liquid oxygen in spite of the drastic temperature difference between the two, with the liquid hydrogen at near absolute zero. I say "extreme" because they had to build a cylinder whose thin walls doubled as a giant fuel tank, in direct contact with the fuel, and as a prime structure strong enough to withstand the stresses of a million pounds of rocket thrust with the weight of the rest of the Stack sitting on top of it—this with materials so light that in combination with the five J-2 engines they would represent only a niggardly 10 percent of the S-II's total weight, the other 90 percent being propellant. Could it be done?

And in 1962 there loomed a long list of other large uncertainties, all affecting design and development. The nature of the Moon's surface, when determined, could upset all prior calculations for the lunar landing module, whichever final form it, in turn, might take. Medical men could not give final answers until after the upcoming Gemini two-man flights, answers on all the human factors involved and which design engineers would have to consider: crew reaction to prolonged weightlessness; lethal radiation; lethal meteoroids; "kitchen" and packaging arrangements for proper diet; hygiene and waste disposal; sleeping arrangements; provision for physical exercise, and many more.

With each new day bringing demands for decisions, Stormy steered his ship through the roiled waters of preliminary design and testing, exuding a confidence which infected his troops, but behind which lay many a sleepless night of lonely doubts. For the thought that drove him, and which he stressed, prophetically, to his subordinates, was this:

"Downey may seem a long way from the launches at the Cape. But it's right here that we've got a life and death responsibility. Don't ever forget it."

BREAKING FRESH GROUND (1963-1964)

"The pleasure of knowing that some new system of yours works well is surpassingly great."

Dr. C. Stark Draper, M.I.T.

If the decision for LOR, and the award just before Christmas to Grumman Aircraft Corporation of the prime contract to build the LM bug, ended one great uncertainty for Stormy Storms and others, it posed a complete unknown to a pair of interested gentlemen at Bethpage, Long Island: Joseph G. Gavin, Jr., Grumman's vice-president for Space programs, and his ingenious design engineer, an Irishman named Thomas J. Kelly, who was to earn the soubriquet of "Mr. LM." The pair had "asked for it" and now they were "stuck with it."

No precedent whatever existed for the creation of a manned spacecraft which would be called upon to function in a vacuum only, and under the pull of only one-sixth of Earth's gravity while at the Moon. On top of that, it would have to be completely self-sustaining, even to the extreme of carrying its own "Launch Complex" with it to a hostile site where no army of ground technicians would be standing by to sweat out days of countdown. On the Moon, it would have to just "up and go" to get home.

Every other segment of the Apollo Stack owed a debt to the technology of long aviation experience in the atmosphere. Hence Grumman could have posted a notice outside its Bethpage plant, "No aerodynamicists need apply." The LM's designers would have to virtually throw away the book in giving birth to man's first pure Space schooner. They had the benefit neither of the Mercury experience and guidelines already at the disposal of Storms' engineers at Downey, nor of helpful clues from unmanned lunar landings, which had not yet taken place. It was all fresh, untrampled snow.

ii

Sitting down to lunch with Gavin and Kelly at the King's Inn Motel, hard by the Manned Spacecraft Center in Houston, I quickly got off on the wrong foot, with informative results, however.

"Would you mind explaining to me," I asked Mr. Gavin, "why your Bug violates the hoary maxim that the more functional a shape is, the more beautiful it becomes? Like an airplane, a seagull in the wind, or a gamefish in the water? Why is the Spider such a monstrosity—aesthetically, that is?"

Joe Gavin smiled good naturedly, with the long-suffering expression of one to whom a question is not new.

"That's easy," he said. "To Tom, here, and to me the LM *is* beautiful."

"He's right," Kelly chimed in emphatically. "Not only that, but it still offends our finer sensibilities to hear disparaging terms like 'Spider' and 'Bug.'"

"We have another frustration, too," Gavin said. "At launch time, when everyone else's hardware is on public display, our light is hidden under a bushel. All you can see is the shell—the adapter section built by North American Rockwell that houses the LM. And then our pride and joy does its thing a quarter of a million miles away from home and never comes back."

"More than that," Kelly added, "we don't really get to see the LM whole even in our own plant, except in mock-up."

"It's either in bits and pieces," Gavin explained, "or shrouded with all kinds of manufacturing and test gear or protective wrappings. But I'd better let Tom do the talking. It's been his baby from the start."

I asked how a relatively small company like Grumman, best known for building fighter planes for the Navy, had beaten the brains out of eight of the aerospace big boys for a prime contract worth nearly a billion and a half dollars.*

"Speculation, on our own money," Kelly said, "with the cooperation of associates who were willing to gamble with us. Back in 1960, when Houbolt first proposed LOR, we decided to get a head start on a proposal for the Lunar Module that we were confident would be needed. We stepped up our efforts after losing the Mercury capsule contract to McDonnell. When NASA finally did call for LM proposals two years later, we were loaded with studies and research data in virgin territory. We'd done our homework, and trained a large stable of pretty knowledgeable gents in a pioneering field, while we were at it."

He paused to dig into a heaping seafood salad and Gavin took over.

"Our problem was simplified," he said, "if anything about the LM program can be called simple, to the extent that we were designing the beast from a standing start, with no preconceptions, that would operate in only one medium—Space. All the other Apollo-Saturn designers had to cope with 'bastard' compromises to make their

*The LM's cost was originally projected at one third of that, for a much lighter and less complex version, however, than was eventually required.

hardware compatible with operations also in the atmosphere."

In my peregrinations about the Space centers, I had been given conflicting answers to a question about the LM's basic design that perplexed me. Why two engines—one for lunar descent and one for ascent—when the larger descent engine could presumably do both jobs? Why not save the weight of the extra engine and its separate propellant tanks and systems? The weight of the landing platform would be left behind in either case. Yet I'd been told that more weight would be saved with two engines and two stages. I related this to Kelly and asked for an answer from the horse's mouth.

"I examined both possibilities very carefully," he said, "since it was one of the earliest decisions I had to make. Weight was not the only consideration—not that much difference either way. But there were a couple of obvious advantages, we concluded, with two engines. First, if you had only one engine for descent and ascent both, and it failed during descent to the Moon, two astronauts would have had it. But with a separate ascent engine, they could fire it up at almost any time before touchdown and safely rejoin the mother ship in orbit."

"Second, on ascent you didn't need a throttleable engine, which was mandatory on descent for adjustments and hovering. Without that requirement, the ascent engine could be made that much simpler hence more reliable, one less feature to go wrong. And God knows you wanted reliability in that ascent engine 'til hell wouldn't have it—in the absence of the backup insurance you had for the descent engine. And there were other advantages. You lifted off the Moon with a smaller, lighter module—only half a spacecraft—for the reunion with the Command Module before its departure from lunar orbit for the final lap home."

Since the LM was the last major contract awarded by NASA, nearly two years after the others, I asked whether this had entailed a backbreaking scramble by Grumman's engineers to catch up.

"Well, I guess we found repeatedly where a man's point of final exhaustion is," he said.

"No exaggeration," Gavin added.

"But in one sense the late start worked for us," Kelly continued. "The state of the art had advanced by that much. For instance, we could use a greatly improved type of battery instead of the fuel cells which North American needed for the larger demands of its Apollo spacecraft. For briefer periods and lower power demands, batteries gave us a weight and cooling advantage. In the end, though, we needed an extra year's time to sweat poundage and even ounces out of our bird; it kept coming up overweight as new requirements arose. We were able to catch up after the Apollo program was delayed after the fire for over a year, even though, because of it, we were saddled with a fresh problem—adding a hundred pounds of fire resistant materials."

I asked Kelly whether Grumman had underestimated the size of the chunk it had bitten off as their program got under way.

"I think everybody did," he answered readily. "Doing any big job really well just plain takes time. Lots more time than you visualize at the beginning. And you're continually wanting to add some highly desirable new feature. So you keep adding weight, as I've said. Then suddenly you're overweight. So you go on a crash diet, redesigning, eliminating what had seemed absolutely essential, cramming four quarts into a three-quart jar. Our final version of the LM evolved fairly closely according to our original design. In concept, that is, but not in weight. The bird grew from a 5,000-pound chick, to a 25,000-pound pullet under the original NASA contract, to a full-grown 32,000-pound hen. Even then we were bursting at the seams."

Our conversation turned inevitably from the past to the immediate present, pregnant with suspense for Gavin and Kelly. As we sat there, Neil Armstrong's crew in

Apollo 11 was well on its way to the Moon. No two men in the country, three days hence, would be watching the first lunar landing more breathlessly on TV. None would feel the impact more personally, nor more dreadfully, should the LM which they had created falter in the clutch. Yet they were able to impart to me their confidence, modest but firm, in a job which they apparently believed in their inner hearts had already been well done.

By the time we arose, my luncheon partners, neither of them large physically, seemed somehow to have gained in height over my first impression. With shoulders to match.

Shoulders wide enough in 1960 to have wielded the pick of the prospector with no guarantee of hitting the pay dirt of government gold. In search of treasure dearer than gold.

Wide enough in 1963 to shoulder the axe of the frontiersman in a new wilderness of Space technology in order to fashion a mobile home on the Moon for two astronauts.

Wide enough to befit that invisible giant in whose hands were soon to repose the lives of Armstrong and Aldrin, carrying out their lonely, rigorous task of mind-numbing precision 240,000 miles removed from the built-in succor of any other hands than those of the giants who had built the LM and God's.

iii

If the LM builders were the last to get off the mark in the Apollo relay race, where batons had to be handed off deftly to contractor teammates in a very literal sense in meshing schedules, how was Doc Draper faring by 1963-1964 up at M.I.T., which had received the earliest NASA prime contract, for Guidance and Navigation?

Draper too had bitten off a larger chunk, relating to the computer, than he or anyone else at his Instrumentation Lab had anticipated—in terms of time and people. Draper had rightly assured Webb that M.I.T. possessed the know-how, from related research on theoretical Mars and lunar flights and from experience on missile guidance systems (most recently the Polaris), to assure the nearly perfect accuracy that Apollo would demand. But the "software," the programming of the computer or "brain" for each of a wide variety of manned and unmanned missions, soon proved to be far more time-consuming, with the involvement of a great many more people, than originally envisioned. Where Alex Kosmala, one of M.I.T.'s wizards of the digital computer, started out trying to keep nearly everything in his own head while programming a relatively simple earlier mission plan, it soon developed that more like a hundred heads would be needed at the lab for later missions. Instead of lending themselves to a standard approach, each programming had to be custom tailored. Help was needed and was forthcoming from subcontractors like SDC (Systems Development Corporation), of Santa Monica, California.

The important thing was that Draper's principles were sound. Yet applying them for Apollo proved to be an enormous burden which fell heavily on the shoulders of Dr. David G. Hoag, the Doc's director for Apollo Guidance and Navigation, Dr. Richard H. Battin, Hoag's head of "Mission Development," and key programmers like Dan Lickly and Alex Kosmala.

In effect, Battin's team was responsible for "flying" every step of an Apollo mission electronically in advance, and storing the navigational information in the memory of the computers built by Raytheon at nearby Waltham, Massachusetts. To put it another way, months before a mission, say Apollo 10, M.I.T. would have to furnish Mission Control, Houston, and the Cape with a computerized simulation of the complete mission, which the astronaut crew would then "fly," perhaps several hundred times, in their Apollo simulators. The real thing would be almost in the nature of a final encore.

In case the foregoing has given the reader an erroneous

impression that M.I.T. "wrote the script" for the missions, it should be pointed out that it was NASA's responsibility, of course, to establish mission goals and flight plans. Contractors TRW and IBM, working closely with Gilruth's headquarters in Houston, provided the planning requirements, which M.I.T. converted into its onboard computer program. IBM's large ground computers, during a mission, furnished the "real time" (namely instantaneous) information needed to correct or update the pre-digested flight plan in Draper's onboard computer. Since the latter's "memory" was 90 percent ineradicable, a permanent and inflexible instruction sheet, any major change in the advance planning of a mission meant that M.I.T. had to resort to an entirely new computer.

In an interview for *SDC Magazine,** Battin frankly recalled the time-bind that M.I.T. found itself in:

> There were a lot of growing pains anyway, of course, but, fortunately, the NASA schedules didn't hold, and the slippage that occurred made it possible for us to do our job. I don't think, no matter how many people we had working on the software, it would have been possible to meet the original schedules.†

Battin's statement does not necessarily conflict with Draper's original promise to NASA that Apollo's Guidance and Navigation would be "ready when you need it." The canny little Missourian could scarely have escaped the conclusion, from a lifetime of dealing with government agencies, that somewhere down the line Apollo would indeed stretch out. As Von Braun used to put it, wryly, you must apply an arbitrary "pi" factor to new programs—raise your sights on estimates of time and cost by a judicious multiple of common sense.

**System Development Corporation* magazine, (August, 1969).

†Alex Kosmala recalls that it took him fifteen months, instead of the scheduled six, to complete the software for the first Apollo guided flight.

iv

In the summer of 1969 I stepped out of a torrential downpour into the musty lobby of a drab, brick, former shoe polish factory on the Thames River in Cambridge, Massachusetts, home of Draper's world famous "I-Lab." Storm clouds from M.I.T.'s adjacent campus were also blackening the skies over the decrepit building, brewed by student militants and by bubble-headed zealots on the faculty, sparked by a professor of linguistics named Dr. Noam Chomsky. The coalition was out to "get" the Doc and his Lab for contaminating a "university atmosphere" with contracts for national defense.

How could this be happening, I wondered as I approached the security desk, to a modest "genius"—so characterized by friend and foe alike—who had given the United States its razor's edge of technological superiority, more than had any other man, over a span of 40 years, for both military and peaceful ends? Why were they singling out the man who was now making it possible to aim Apollo 11 toward the Moon with an error of less than a small fraction of one degree in direction or one-tenth of one percent in speed? (This would be the comparatively elementary requirement to insure a hit by a projectile anywhere on the huge target of the Moon; Draper would have to "miss" the Moon intentionally and score a bull's-eye 70 miles to one side for a lunar orbit.)

Waiting for the escort required by regulations to take me to Draper's fourth floor office, I speculated whether the Doc could possibly be as unpretentious as the legend. My answer was promptly forthcoming when the escort materialized in the miniaturized person of Draper himself. He signed my clearance, openly flirting the while with the curvy young miss behind the reception desk. With the green eyeshade he was wearing, in shirt-sleeves, he could have been mistaken for the janitor. Except for a remarkable pair of eyes. You saw wisdom there at first glance, that and combativeness. You felt that he was ready for a fight or a frolic at the drop of a hat. Or a test tube.

As soon as we entered his cluttered office, he shed his shoes and padded about thereafter in wool-sock comfort as we talked, the first topic being the anti-Draper campaign.

"How long do you expect to remain in business here?" I asked without any preliminaries. He shot a pained look at me.

"Do you think you're kidding?" he said. "They mean real trouble. I'm not about to quit, though, they'll have to drive me out." Then he added with quiet bitterness:

"You know, I never realized that I was a warmonger."

Apollo 10 being at that moment more than halfway to the Moon, I asked him how many of the three scheduled IMU alignments had been necessary, referring to the midcourse corrections which could be fed into the Draper Inertial Measurement Unit if need be. He grabbed his phone and called Mission Control in Houston, receiving an immediate report, then turned to me beaming with satisfaction.

"They just skipped one," he said. "The crew's optical sighting on the Earth's horizon showed the trajectory is accurate *without* an alignment."

Not only in his ready answers to me, but in the way he accepted interruptions from anyone who cared to wander in, answering his own phone the while, he showed himself to be about as approachable as a man can get. I am sure that if a mail-boy were to stop him in the hall and ask him a naive question about gyros, he would take the pains to enlighten him on the spot.

He is utterly frank. I could see where his bluntness has raised the blood pressure of prominent people. He did not hesitate to characterize as "junk" any navigational hardware developed by others that did not measure up to his idea of accuracy, namely that one should think in terms of feet, not a mile or two, when speaking of long-range errors.

Except during our brief allusion to unrest on the campus, he was brimming with cheerfulness and good humor. And well he might be. The latest and greatest of his brain children was working, and better than anyone else had dared to foresee. By his own definition of technology as the activity by which man purposefully alters his environment for desired ends, he had fulfilled himself. More than a brilliant scientist, more than a practical engineer, although he was both, he had crowned his career by enlisting these two disciplines in the further professional maturity of the true "technologist"—that scarce species whose training enables them to plan, manage, and *achieve* programs that give humanity wider choices than mere adaptation to or acquiescence in its natural surroundings. The Doc has little patience with those who restrict themselves to theorizing, specializing, or educating themselves from books with "answers in the back," without ever really addressing themselves to original solutions to society's needs or performing a responsible *service.*

By way of a postscript, Dr. Draper was finally dethroned in January 1970 and has departed M.I.T., in as shabby a reward for contributions to the safety and prestige of one's country as that bestowed upon Sir Winston Churchill by an electorate which ousted him as Prime Minister, once victory over Hitler's Reich had been achieved.

V

While other contractors were still getting up a head of steam in 1963, one company was racing toward the finish line, which it crossed the following year, far ahead of the pack. And like the holder of a life insurance policy, it was praying that its product would never be needed.

The Lockheed Propulsion Company, Redlands, California, had been given the responsibility by North American for delivery of the two solid rocket motors in the Launch Escape System that could catapult and pitch over the Apollo spacecraft into a safe abort at any time from pre-liftoff through the first three minutes of climb-out. Thiokol supplied a third motor at the top of the North American-built miniature Eiffel Tower for routine separation and jettisoning of the entire insurance package as soon as its three-minute coverage had elapsed.

"How did you win the contract?" I asked Irwin "Irv" Spitzer, the personable young manager of Lockheed's LES program.

"Brevity and price," he said. "Our proposal to North American was only twenty pages and our price tag was low, in spite of which we *under*ran it."

"That's a new twist," I said.

"Another advantage we had over the other bidders, Aerojet and Thiokol, which is formidable competition, was our experience with the LES for Mercury."

"You missed out on Gemini?"

"No, Gemini didn't use a launch escape tower like Mercury and Apollo. The two astronauts had ejection seats, like in a jet fighter, and came down under personal parachutes. Entirely different system."

It was both unusual and refreshing to encounter, in Spitzer, a program manager who did not have a tale to tell of heartbreaking adversity. It almost appeared that he had enjoyed, rather than fought, problems, if indeed he considered any challenge a "problem."

For example, the incipient problem of steering the LES tower and the astronaut's capsule away from the Stack thundering on its heels, by the use of vanes or gimbaling the engine, with penalties in weight and complexity, was circumvented. Thrust Vector Control, as the requirement was labeled, was achieved by a ridiculously simple solution: design the four-nozzle motor so that the bird couldn't fly straight under any circumstances. This was done by making one nozzle a little smaller, and the nozzle opposite to it a little larger, than the other two nozzles. Thus the bird flew automatically in a curving arc out of the Stack's path, like an arrow with a bent feather.

Under close questioning, Spitzer cheerfully admitted that his back was covered with figurative welts from the lash of urgency. North American needed Lockheed's LES at the earliest possible moment for abort tests of its boilerplate models of the Apollo capsule, before it could proceed with production of flyable models.

Without smugness, but certainly with contentment, Spitzer looked back on a program that came out on time, below cost, with the first Apollo system to be certified man-rated. It was all right with him, too, that nothing could go wrong with a system if it was never fired, and so it was possible for him and his team to pick up their tools three minutes after launch and go home.

It pained Spitzer, however, that one of his young sons did not seem to appreciate the extent of his father's responsibilities.

"On one mission, just as the countdown reached T-minus 8 seconds," he told me, "my son asked if he could switch channels to his favorite TV cartoon program."

In parting I asked him how he would feel about accepting an assignment to another task of Apollo-like proportions.

"No problem," he answered promptly.

vi

Working hand in glove with "Irv" Spitzer and, of course, Storms' Space Division, on Apollo's Launch Escape System, were the parachute builders of Northrop's Ventura Division, in nearby Thousand Oaks, just west of the San Fernando Valley.

"Now it's your turn," Spitzer could have said to Northrop's closeknit team of Wesley A. Steyer, program manager, and his "paradynamic" specialists William H. Freeman and Theodore W. Knacke, a German who had worked for Von Braun. If the escape tower had been called upon to fire, Lockheed's baton would immediately have been handed off to Northrop and its three huge landing parachutes.

Like Lockheed, Northrop would have to hold its breath from the time the astronauts entered their capsule until the LES was jettisoned. Unlike Lockheed, it would have to continue to hold its breath for eight days until the very last moments of the mission, after Apollo's fiery reentry, and until the capsule plopped safely into the Pacific. For either event, and it was bound to be one or the other with the same set of chutes, Steyer had to provide Storms and Spitzer with a product that was more than 99 percent reliable. And this reliability was supposed to be achieved with a product that, at least in the public's mind, had demonstrated itself over the years to be notoriously *un*reliable—the parachute.

I have seldom met two gentlemen who were so utterly—perhaps fanatically is not too strong a word—fascinated with their lifework. When Freeman, a pioneer for the Navy in parachute development and test work, was talking, Knacke could scarcely contain his eagerness to leap into the breach and help enlighten an author with an expressed interest in the forgotten men of Apollo. And vice versa. It was as if at last they were getting "equal time" with the other builders. Both experts were excitingly articulate. Their enthusiasm was contagious. Knacke's words tumbled out so fast in his heavy German accent that he was not easy to follow. Freeman took the floor first.

"Before we got started on the Space program," he said, "parachutes were considered one of the 'Black Arts.' There was a dearth of precise data. We didn't really have the foggiest idea of reliability for a system of the scope and complexity of Apollo. The Army, for example, had been airdropping trucks and heavy equipment under clusters of so many chutes that even if half of them failed, there was enough margin for the rest of the chutes to get the job done. You could get by with that kind of crude dependability where weight was no big problem. But the first thing we had to do from Mercury on, where every ounce and even gram of weight was a serious matter, was to vastly improve our store of test information. We did this by attaching miniature sensoring devices to vital stress points in the riser lines and nylon panels, so that we could measure the strains accurately, for the first time, during flight tests."

"May I interrupt," I asked. "As a pilot I've had two emergency jumps and I was under the impression that my chute was pretty reliable. For one thing, they were drop-tested periodically with dummies in the Thirties, when I was at Langley Field."

"Let me answer that," Knacke broke in, "in two parts, please. Point one, the reliability question involved in a personal parachute for a pilot weighing, say, 200 pounds, is quite rudimentary compared to an Earth Landing System for a Space capsule weighing, in the case of Apollo, several tons, with drogue chutes and then pilot chutes opening at high velocities before the main panels open, and with all that gear crammed into a minimum of storage space. Point two, there has been a gross misconception about chutes presumably failing to open."

"It's a canard," Freeman hastened to agree. "Chutes *don't* fail to open if the pilot pulls the ripcord."

"I'll explain," Knacke continued. "You read in the newspaper that the pilot's chute failed to open. What really happened? Why he failed to pull the ripcord, because in abandoning a disabled aircraft he hit something that knocked him unconscious or killed him. Or he froze at high altitude before it was safe to deploy the chute, or was otherwise incapacitated."

"I can cite you one year," Freeman joined in, "in which there were 32 deaths among paratroopers out of 35,000 drops—*none* of them due to chute failure. The deaths were caused by drowning, being dragged in high winds, collisions with trees, fences, buildings, and other obstructions—accidents like that—and human errors connected with bailing out of the aircraft."

With this background out of the way, Freeman and

Knacke touched on some highlights of the predicament they found themselves in when Storms allowed them just 485 pounds for a total package that would provide a failproof system for the lowering of a six-and-a-half ton spacecraft to a gentle landing from any altitude below 40,000 feet down to the low altitude of a pad abort. Through no fault of his own, Storms found himself perforce cast in the role, from Northrop's viewpoint, of a tormentor of sadistic cunning.

Three main panels would be required, the third being for insurance. Any two panels would provide a safe descent, the odds against failure of more than one panel being astronomical. But it looked at first as though it would be literally impossible to stow those three ring-sail type chutes in the narrow confines of the "collar" around the neck of Apollo's cone. Each chute had a diameter of 85½ feet, enough for a fair-sized circus tent. There just plain wasn't enough room and there was no way more space could be permitted. There was only one way to go. Since you couldn't expand, though NASA's requirements kept on expanding, you had to contract. Northrop's solution borders on black magic.

If you've ever stuffed cotton into a cigarette lighter and been amazed how much more you could keep forcing into a tiny compartment that seemed already full, you have some idea, on a small scale, of Northrop's answer. Its engineers crushed the folded nylon panels under presses at 300 pounds per square inch in a vacuum (they even had to squeeze all of the air out of the fibers) until the molecules were compressed to the density of wood.

"We wound up having to compress the folds to the density of the top of this maplewood table," Freeman said. "No exaggeration."

"But how could the material *un*fold again," I asked incredulously, "when the chute was deployed?"

"Because of nylon's great elasticity," he said, "it holds its shape surprisingly well. In one test, after the chute had been packed in its bin for two years, the nylon didn't bounce back by itself immediately—remained creased—but it regained its elasticity after it had been unfolded."

"It seemed that every time we'd succeeded in cramming our chutes into the available space," Knacke went on, "NASA would turn the screws on our thumbs again, with demands for more panel area at less weight. It was like trying to thread a needle with a rope, but somehow we'd come up with more strength in lighter fibers, compressed still more."

The lowly fishline provided the solution to a second formidable problem that faced Northrop. Air turbulence in the immediate wake of the descending spacecraft dictated that the risers to the parachute canopies be long enough to separate the chute from the turbulence and keep it stable. However, in tests the chutes were proving to be anything but stable.

"Max Faget was giving us hell," Knacke said, "because he thought our design must be wrong. But the real trouble was that those long lines were flexing, sort of yo-yoing. We needed material that wouldn't stretch. Then someone had a brainstorm, recalling a type of fishing line that *didn't* stretch. Sure enough, we were able to adapt the fishing line, in the form of a mesh, to shorten the lines and solve the flexing that was causing the parachute instability."

The reliability "barrier" forced Northrop to the extreme of individually inspecting every one of *millions* of stitches in the chutes. Even so, Apollo landed with holes in the main panels the size of baseballs, the result of friction's effect on the fibers. This was allowable, since any increase in the rate of descent turned out to be negligible.

Electronic devices had to guarantee an order of precision appropriate to the simultaneous opening of the two drogue chutes, followed by the opening of the three main panels within one-tenth of a second of each other. It is easy to visualize what would happen if the three mains did not open at almost the same instant. The first chute to open would momentarily absorb the entire shock of Apollo's weight and could tear itself to shreds. Then there'd be no backup if a second panel failed too.

I doubt that anyone who had accompanied me would have left Thousand Oaks without some feeling of awe for Northrop's Ventura Division. Its engineers had been told to "jam forty Chinamen into a taxicab," as Freeman put it. And they had done it.

The visit brought to my mind a quote attributed to that famous producer, Samuel Goldwyn, when he was confronted by an actress with an outrageous salary demand for her next picture.

"I can answer you in two words," Sam said.

"*Im* . . . possible!"

THE HALF-WAY FLAG (1965-1966)

"A Space effort is in some ways almost as uncomfortable to live with as a war effort."

Tom Alexander, Science Reporter

In sharp contrast to the complacency, or apathy, with which most of us in 1970, before the ill-starred Apollo 13 mission, were inclined to accept the *fait accompli* of Americans sojourning on the Moon (the author was not immune to the complacency), loomed the tall, treacherous mountains of technological difficulties which seemed to threaten NASA and Apollo's prime contractor in the middle Sixties with failure to reach the common goal.

NASA was becoming increasingly alarmed by the record of progress at North American, where schedules for the Apollo spacecraft were slipping and costs were running over original estimates at Storms' division in Downey. The same was true, and even more so, with the development of the company's Saturn S-II stage, also Storms' responsibility, at Seal Beach. Accordingly, Major General Samuel C. Phillips was dispatched by NASA from his Washington office to scrutinize North American's apparent shortcomings and recommend some quick answers.

The "Notes" which Phillips and his team compiled and submitted to Atwood in December 1965 were to receive wide publicity after the fire on the pad approximately a year later, as the controversial "Phillips Report." (North American's files at the time did refer to "Notes," which became a semantical point of contention in 1967 when Webb and Atwood were questioned about it.)

Before one passes judgment on the events about to be related, it has to be kept clearly in mind, with the benefit of hindsight, how drastic was the sheer magnitude of what had to be attempted in so short a time. I see an adjusted perspective, rather than exaggeration, in a recent statement by noted aerospace writer James J. Haggerty:

"So extraordinary were the demands for performance and reliability needed to land men on the moon that the Apollo team had to create an entirely new order of technology and to compress *several decades* [italics mine] of normal technological gain into less than one. Advances in *aerospace* technology were not, by themselves, sufficient for the task; it became necessary to force progress in virtually *every* scientific and technological discipline."

Lee Atwood and Stormy Storms found themselves at the center of the mandate to compress time. They could not ask, as many critics have demanded then and since, "Why all the bloody rush?" The goal had been set and it had to be met—perhaps as early as 1967, if all went well.

All did not go well. It could not possibly have.

How could all go as scheduled when, if you were Storms, you were breaking new trails every day, hacking your way with a mental machete through the tangled vines of a designer's jungle?

When there was no way, for openers, to duplicate the "hard" vacuum of Space, even in the newest test chambers, for the design and development of Apollo systems that had to be 99 percent reliable for mission success and

99.99 percent reliable for safety, in the real vacuum of Space?

When you did not know until your program was nearly three years along whether to design for a water landing or for the much harder impact of a descent to a land surface? Water recovery was not chosen as the primary mode by NASA until November 1964.

When costs and scheduling, scarcely a precise art in the mass production of automobiles, for example, were subject to those continual changes that are inseparable from even a relatively simple program for a new aircraft? Apollo's intricacy was several orders of magnitude greater than that of an SST (supersonic transport). No one had ever designed a vehicle to carry men to the Moon and back.

When you were so harassed by crises in trying to meet cost and schedule obligations that it was a wonder there were enough hours left in any day to manage the rest of the program?

When even the finest human brains and the most robust constitutions cried out for relief from near-exhaustion, only to be saddled with fresh and heavier demands?

When, to cite one more example out of an endless number, you found yourself stumped by the challenge of so presumably rudimentary a device as a satisfactory fuel gauge? Motorists and aircraft pilots might ask, "Aren't fuel gauges old stuff?"

Far from it, when you're talking about spacecraft where your liquid fuels will be floating around in a weightless condition. Space Division was able to develop acceptable measuring devices in the form of probes which sensed the quantity of fuel in the tanks electronically, but only after a rather far-out technical detour with a "nucleonic concept" of using radioactive particle emissions "to measure irregular and unpredictably shaped masses in a zero-G environment." Try that mouthful on your auto mechanic.

ii

To liken the stresses and painful progression of Apollo to a war effort is a fair analogy in more ways than one. There is the parallel of America's bomber force in England in World War II. In spite of the momentum of every resource of a great nation behind Generals Eaker, Spaatz, and Doolittle, successively, in the building up of the aircraft and crews for the Eighth Air Force, skeptics had a field day deploring the absence of impressive results during the first three years of daylight precision bombing. What the critics overlooked, as they were to do also during Apollo's formative stage, was the length of the gestation period. The war was approaching its climax by the end of the time it took to build the Eighth Air Force into a full-fledged bombing weapon system, following which Germany was brought speedily to its knees.

So with Apollo. The payoff matured with a rush of initial successes that tended to erase from people's minds the earlier agonies of creation. When it finally occurred, the lunar landing was akin for NASA and North American to a three-foot putt for a golfer who has spent most of his time in the rough before reaching the green.

With the foregoing in mind, the letter that Phillips wrote to Atwood, dated December 19, 1965, was fairly predictable and is quoted here in full:

> Dear Lee:
>
> I believe that I and the team that worked with me were able to examine the Apollo Spacecraft and S-II stage programs at your Space and Information systems Division in sufficient detail during our recent visits to formulate a reasonably accurate assessment of the current situation concerning these two programs.
>
> I am definitely not satisfied with the progress and outlook of either program and am convinced that the right actions now can result in substantial improvement of position in both programs in the relatively near future.
>
> Inclosed are ten copies of the notes which we compiled

> on the basis of our visits. They include details not discussed in our briefing and are provided for your consideration and use.
>
> The conclusions expressed in our briefing and notes are critical. Even with due consideration of hopeful signs, I could not find a substantive basis for *confidence in future performance* [italics mine]. I believe that a task group drawn from North American Aviation at large could rather quickly verify the substance of our conclusions, and might be useful to you in setting the course for improvements.
>
> The gravity of the situation compels me to ask that you let me know, by the end of January if possible, the actions you propose to take. If I can assist in any way, please let me know.
>
> Sincerely,
>
> (signed)
> Samuel C. Phillips
> Major General, USAF
> Apollo Program Director

Even couched as they were in restrained official rhetoric, these were harsh words indeed. Atwood was being told that his company was falling down on the job so badly that NASA had lost confidence in its ability to deliver the Apollo goods.

Atwood's reaction was to comply to the hilt with Phillips' suggestions. He instituted an immediate company study by an army of top drawer survey teams under Harvard W. Powell and C. Wesley Schott. The reply to NASA which resulted was constructive.

Officially, the company agreed with NASA that improvements could and should be made in management, more efficient utilization of personnel and closer control of costs and schedules from topside; at the same time, Atwood made clear his belief that many appropriate corrective actions had already been set in motion prior to Phillips' survey, and that most of Space Division's difficulties stemmed from the constant changes and the hidden reefs that afflict all new large-scale endeavors.

Privately, the emotional and subjective reaction at

higher levels was more like, "We don't for a moment concede that our company has forgotten, overnight, everything it's learned over a span of 26 years in running large programs. Proven performers haven't suddenly become bunglers. We're committed to giving the astronauts the best spacecraft—the finest chunk of machinery ever built—and by God we're going to do just that."

iii

All this ferment, of course, was very close to home for Storms. Already well known for extracting the last drop of effort from his troops, he redoubled his campaign of persuasion, when necessary, to the point of psychological warfare.

"If you can't cut it," he told a conscientious, able subordinate who was having nightmares coping with the latest problem on the S-II booster, "I'll send somebody over there who can."

"You do that, Stormy," the subordinate retorted, angry and undaunted, "if you think you can find somebody better. I've had it up to here. Go right ahead."

The anger was what Stormy was hoping for. He shifted to an entirely different tone, grinning.

"Can't you take a little needling?" he asked. "If I didn't have the sense to know you're the best in the business, *I* should be fired."

Having delivered his compliment the hard way, Stormy sat back and watched the engineer knock his brains out until the problem was licked. This kind of loyal response beyond the call of duty which Storms was often able to evoke is reminiscent of the early days under "Dutch" Kindelberger, when North American was struggling to make ends meet on an airplane contract. The legend has it that a manager came upon a German engineer working his men after hours to finish an urgent job.

"Man, you must be crazy," he said. "You know we can't pay overtime!"

"Is none of your business vot vee do on our *own* time," the German answered. "This vee do for the *Dutchman.*"

When the panic button has been pushed—and Space Division's searching self-examination following Phillips' displeasure was not without overtones of panic—any organization is lucky to have some cool heads. One of these belonged to a horse-sense engineer named Charles H. Feltz, who was the durable number two man to Dale Myers, program manager for the Apollo spacecraft. "Charlie" never lost sight of the practical and the attainable amidst all the redoubled strivings at Downey for near-perfection. In a nutshell, his philosophy is expressed in a quote from his earlier, simpler career in building airplanes:

"Does it fit?" he asked a thwarted engineer.

"No, Charlie."

"Does it *touch*?"

"Yes."

"Then, godammit, nail it down."

iv

It is beyond the scope of a nontechnical book of this nature to try to "nail down" the intricacies of the literally millions of functional parts that had to be brought together, against exacting deadlines, in the creation at Downey of the Command and Service Modules (the CSM). Nor is it really necessary, perhaps, to describe the spacecraft's characteristics in detail, since everyone who has owned a TV set during the past few years has been briefed on such specifics 'til hell won't have it. I'll attempt only to convey some broader interpretations of the kind of near-miracle being stalked by fallible big game hunters, much as the Lilliputians stalked the giant Gulliver.

Look at it this way. Although the very idea of going to the Moon was no longer synonymous with lunacy, how much more brash was the undertaking to introduce into the heavens Man's own tiny version of his own planet Earth in order to accomplish it? True, the new space body would have to provide life support for only three rather than three billion inhabitants, but it must be built to harmonize with the same natural laws that govern other celestial bodies, and in at least one crucial respect it must go the Creator of the Earth one better. It must have that characteristic that distinguishes animals from plants. Mobility. Other heavenly bodies move, of course, but only in rotation about their own axes or in fixed orbits. Storms had to build a miniature Earth that could rove around the local area of our Universe as its driver pleased, while enjoying a comfortable atmosphere, clement temperatures, plenty of food and water, and reasonable protection from lethal radiation and meteorites. Measured against the task, a modicum of human errors perpetrated along the way by Apollo's builders must be accepted realistically. The surprising thing is that there were not more of them.

Designwise, an odd coincidence may be worth mention. After all of the scientific geniuses got through determining the shape of the "home" in which the astronauts would journey Moonward, after all of the studies and computer analyses and tests, what did they come up with?

A wigwam.

The conical capsule which shelters Apollo's crew bears a strong resemblance, even in its diameter of approximately 13 feet at the base and its height of nearly 11 feet, to the primitive tepees in which the American Indian sought protection from the elements.

V

During its fabrication at Downey under the able team of Storms, Paup, Myers,* Feltz, and Ryker, which came to include George Jeffs, Mike Vucelic, Bob Fields, Gerry

*Dale Myers was not freed of his full-time duties on the Hound Dog missile program until 1964.

Fagan, George Merrick, Dave Levine, Gary Osbon, Bud Benner, and many other star performers, the "wigwam" evolved through much the same stages as a new aircraft.

First, the many mockups, those full-scale models in which dimensions but not materials are important.

Second, the boilerplates or "dummies," not completed in all respects, but made of materials suitable for realistic testing, such as drop tests on land and water, and vibration tests.

Third, the Block I capsules, both unmanned and man-rated for Earth orbital missions; and fourth, the Block II versions which would fly men to the Moon. The distinction between Block I and Block II spacecraft became pertinent when White, Chaffee, and Grissom perished on Pad 34. The crew was testing a Block I version, less advanced and more in the nature of a prototype than the Block II versions which supplanted them, and which, it should be noted in all fairness to the builders, incorporated many improvements by North American as it progressed up the "learning curve." For example, better design and protection for the wiring in the cabin, most likely source of the fire, had already been incorporated in the Block II spacecraft prior to the tragedy.

Not easily visible to the layman was the enormous scope of the testing that accompanied Apollo's gestation period. Up to 1966, Storms' builders delivered eighteen mockups and twenty boilerplates (besides two finished spacecraft) to NASA for the proving out of systems and components. In addition to countless ground tests, samples of the kind of flight testing that necessitated so large an output of preliminary hardware included:

Pad abort: Tests of the launch escape system's ability to work in an emergency before launch while on the pad. Successful.

Transonic abort tests: Little Joe II boosters were used to simulate a Saturn V in trouble in a high stress, high speed region. Successful.

Tests of spacecraft compatibility with the Saturn I rocket, in Earth orbit. Successful.

High-speed abort tests: Verification of the launch escape and earth landing systems. Successful.

Nearly one-half the total effort that went into the delivery of a finished Apollo spacecraft, I was told, lay in testing, as distinguished from design, development, and manufacture. The nearly endless repetitiveness of that testing, as much perspiration as inspiration, reminds one of the thousands of balls a champion golfer or tennis player must hit in practice before he can achieve the order of reliability in his game that is often mistaken for genius. One illustration that sticks in my mind: it took six *months* at Downey for the final checkout of a completed spacecraft on an elaborate test stand—this after each part, component, and subsystem had already been subjected to rigorous tests separately, beginning with the smallest vendor, then the subcontractor, and thereafter at successive stages of fabrication—to be sure that every system was integrated with the whole. Then, after delivery to the customer, of course, more tests continued right up to the moment of launch at Cape Kennedy.

Toiling shoulder to shoulder with North American in the overall test program were NASA and Storms' major subcontractors such as Honeywell for stabilization control, Collins Radio for communications, AirResearch division of Garrett Corporation for environmental control and, as has been described in some detail, Northrop and Lockheed for the abort and earth landing systems combination.

A haunting question still remained for Storms, and at NASA for Sam Phillips and Joe Shea (the responsible manager at Houston), despite all of the concerted striving for near-perfect reliability: "Is there something we've overlooked?"

They were not to receive an answer until the ill-fated year of 1967, to be dealt with in the next chapter. Meanwhile, when Storms could wrench his attention away from the Apollo spacecraft, he had plenty of cause for concern over at Seal Beach on the progress, or lack of it, on the S-II stage for the Saturn V booster. There, in a change of top management, an individual of unusual stature, both in physical height and competence, had been placed in charge—a tall drink of water, and retired Air Force general, named Robert E. Greer.

vi

Riding in a small company helicopter above the sprawling oil refineries and harbor of Long Beach, en route to my visit with Greer at nearby Seal Beach, I found my interest centered less on the headaches that had beset the S-II program than on the man who had been selected to solve them. Not that the headaches, technically intriguing and formidable enough, had lacked drama. The success of Apollo in reaching Earth orbital speed hinged on the performance of Greer's huge S-II second stage rocket booster. It was that an opportunity lay before me to learn something perhaps more intriguing about the human equation.

Why Robert E. Greer? A former West Pointer, presumably limited by the rigidity of the "military mind," he was also a sometime schoolmaster who had taught electronics at the Military Academy. Diverse flying duties later led to his rating as a Command Pilot and the rank of major general, with large early responsibilities in missile and Space programs for the Air Force which came to fruition under Benny Schriever. Why, for that matter, a man with no fancy engineering degrees, only a B.S. and a B.A., and with no management experience in private industry?

And yet he had been one of only two men (Sam Phillips being the other) who had received a personal endorsement of straight "A's" across the board from every single knowledgeable person I interviewed in the Space program—in NASA, in the military, in industry, and in the scientific community.

I sniffed, along with the acrid aroma of crude oil refineries from below, another clue to Neil Armstrong's first footprint on the Moon.

Certainly the sudden choice of Greer in January 1966 was accelerated by the disturbing warnings of Phillips' survey task force the previous month. Atwood had promptly delegated larger responsibilities to veteran manager Ralph Ruud, his good right arm on the aircraft manufacturing side of the house, to oversee both the Apollo spacecraft and the S-II programs, separating the two into two "companies," so to speak, under Storms, with Dale Myers exercising greater autonomy for the spacecraft's production control, and with comparably greater autonomy to be given to a new manager of the S-II program.

Greer, who had only recently joined the company as an assistant to Storms, became that man, succeeding William F. Parker. In an unusual arrangement, and one commonly foredoomed to failure, Parker stayed on as Greer's subordinate. As it turned out, Bill Parker provided helpful continuity, serving loyally, effectively, and without rancor.

The chopper skimmed over a tall S-II assembly building, settled on the grass in front of Greer's headquarters, and I was whisked to his sanctum on an upper floor, pencil and pad at the ready.

vii

The first thing that struck me about Bob Greer was the absence of any feeling of haste or pressure while I was with him. It was as if he had all the time in the world, although I knew better. I happened to have picked a day when the conference rooms outside his modest office were buzzing with visiting firemen from other companies which were teamed with Space Division on its upcoming proposal to NASA for a post-Apollo Space station. The heat was on. Yet he was unhurried and courteous throughout the four hours we spent together, with a knack of leavening even the most serious discussion with humor. His responsibility he took seriously. But not his title.

"Let's go eat," was his first and welcome pronouncement.

What I've said about the absence of haste does not apply to the long strides he took down the hall on our way to the elevators. He's a fast walker. I had plenty of exercise in store, virtually running to keep abreast of those long legs.

During a leisurely lunch at the adjacent Old Ranch Country Club (I joined him in a single beer in lieu of the cocktail he offered), he gave me some of his management philosophy.

"It was important for any ex-military officer like me to lean over backwards not to resemble an arbitrary type of manager with civilians . . . which isn't my natural approach in any case, in or out of the military. I try to suggest rather than give orders—up to the point, that is, of shouldering clear responsibility for final decisions. That you can't duck."

"People tend to get polarized by function, with engineers fighting production troops, and production men fighting quality control types, and so on. I've tried to break that up by frequent meetings of all parties at which we can bear down on the overall, common objective . . . see the forest instead of one man's grove of trees."

"When a program's in trouble, it can be self-defeating to beef up your organization with more people. What's often needed is fewer people, but with the right answers. Two heads can be better than two hundred."

"If I'm forced to take sides, I'm a champion of the quality control guys. I'd rather let engineering and production scream their heads off about cost and schedule overruns they blame on holdups by quality control. Because the reverse, overemphasis on making engineering and production happy about costs and deliveries can insure a poor product. And that's what's *really* expensive."

Our first stop after lunch, at Greer's control room, was reminiscent of other control rooms I've seen at Air Force commands. "Visibility" for management it's now called. Continually updated displays on the walls provide an X ray of all the important "bones" of the operation—where it's ahead of schedule, where it's behind, and where the trouble spots are. And another aspect of Greer's service background was now paying dividends—the tradition that "you never stop going to school."

Greer had supplemented his West Point education by postgraduate courses in electronics, math, electrical engineering, metallurgy, and atomic physics as his career led him into new fields of aerospace such as missilery and inertial guidance. More recently, he had boned up on the esoteric science of fracture mechanics, highly pertinent to an understanding of the defects, porosity problems in welding, and structural stresses peculiar to the fabrication of the S-II.

"I had to know enough so that the experts couldn't snow me," was the way he put it.

Still another reflection of Greer's military background was his insistence on more orderliness at Seal Beach, if necessary at the expense of speed in getting things done.

"Your people have got to know from day to day where they stand," he explained. "You can't keep shifting the ground under them. Orderliness in administration can be a curse if it degenerates into red tape—a cause to be worshipped for its own sake—but there's been too much of a hell-for-leather approach sometimes in our so-called crash programs. You can't say, in effect:

'Come on, fellows, we're gonna build ourselves a rocket, let's go.'"

Often the boss will delegate a star subordinate to take you on a VIP tour of his operation. Not Greer. Or perhaps, from his early experience as a teacher, and knowledge of his subject, he *was* the star briefer at Seal Beach. No specialist accompanied us to answer the tough questions, as had been my experience on many an occasion, nor did he fumble or evade any questions. A man of deep, quiet confidence in himself, devoid of false modesty without being vain, he inspires confidence in others.

When we walked over to the assembly building, I was forcibly reminded once more of the sheer size of the Apollo-Saturn rocket. Looking at it fully assembled on the pad at the Cape, slender and symmetrical, you get nowhere near the same conception of its magnitude as you do looking at one of its major parts, like the S-II, separately. Standing 81½ feet tall and 33 feet in diameter, it is an enormous rocket in itself, yet it is merely the smaller stage that fires when the mighty Saturn 1C first stage on which it rests has dropped away toward the ocean. It's so big that it had to be shipped from Seal Beach by barge through the Panama Canal to the Mississippi Test Facility at Bay St. Louis, where Von Braun's men testfire the complete S-II bird for the first time.

Essentially, the S-II is a large fuel tank to feed its five J-2 liquid hydrogen engines. So what's the big deal, one might ask, about building a round, aluminum "can"? Greer explained why it's not so simple.

That old nemesis, weight, had been the big bugaboo. Inflexible limitations on weight, as in other segments of Apollo if it were to fly at all, dictated the thinnest possible skin for the shell—so thin that they were right near the borderline of sufficient structural strength, with almost no margin of safety—so thin a workman had been able to hear a washer the size of a dime sliding around in its cavernous interior while it was being transported from assembly to dockside. Greer's predecessors had found out where the safety margin was the hard way when they filled an early S-II vehicle with water in a maximum test of the strength of the primary structure. It burst at the welded seams and disintegrated like a paper bag full of water dropped from a window.*

*The test was not a "failure." Normal limits were actually exceeded.

The other chief culprit was the hydrogen fuel. It doesn't want to be a liquid. It wants to be a gas. To keep it liquefied means supercold temperatures (cryogenic), and supercold temperatures create a problem when there is a thin common bulkhead separating the liquid hydrogen from the liquid oxygen. The latter is also supercold, but warmer than the hydrogen, enough so to cause the liquid hydrogen to boil into a gas in the absence of proper insulation. Ergo, new near-miracles had to be wrought in insulation, and were. Even so, the S-II's builders might never have "made the weight" except for a helpful byproduct of extremely low temperature. It strengthens the molecular structure of the S-II's tanks, which are in direct contact with the liquid hydrogen (the insulation is applied to the exterior of the tanks). Thus the metal skin need not be as thick, hence as heavy, as it is on Douglas' similar but smaller S-IVB third stage, in which the insulation is on the *inside.* Douglas engineers faced the same weight and skin-thickness problem as North American, but faced it earlier in the Apollo program (for earlier versions of the Saturn series of boosters), before bonding technology had advanced far enough to permit exterior insulation.

Under Greer's aegis, the team at Seal Beach had moved steadily ahead on the course that would enable it to meet its commitments on time. A humorist has said that it takes a lot of hard work and dedicated people to accomplish the inevitable. I left Bob Greer with a clearer idea of why the success of the S-II, or any other program in Apollo, was in no sense inevitable.

A significant ingredient of that success had germinated way back in the Twenties when a handful of Air Corps "rebels," inspired by General Billy Mitchell, threw away the conventional book on military strategy in their fight for air power. They thought big. They kept their minds wide open to the new, without closing their minds to proven axioms. Thus they infected a new generation of Air Force leaders in the explosive expansion of World

War II with a new kind of "military" mind—a mind broad, flexible, and imaginative enough to see us through a new and terrible kind of war, the military nightmares and complexities of the nuclear age, and spearhead our ventures into Space, both military and peaceful.

Greer has been spotlighted here, at the risk of exaggerating his role, not to canonize him, but to let him serve as a representative example of a breed which includes many other Air Force dynamos who have contributed drive to NASA's civilian Apollo effort. Men of the stripe of Sam Phillips, Carroll Bolender, Ed O'Connor at Huntsville, Rocco Petrone* and "Buz" Hello at the Cape. Or "Deke" Slayton and Eugene Kranz, vital cogs in flight operations at Houston. And, of course, many of the astronauts. Appropriately enough, NASA's top man, Jim Webb, belongs in the club; although largely overlooked, his credentials (and he is especially proud of them) include the wings of a pilot in the Marine Corps.

These men, and I have mentioned only a small sampling, have been a product of disciplines, nurtured in the adventurous hazards and mechanical complexities, first of aviation and then of aerospace, totally foreign to the comprehension of a J. D. Salinger, a Philip Roth, or an Arthur Schlesinger, Jr., who would have no conception of the kind of shared pride in the fraternity which sustains them.

In other words, these men are "squares," esteemed with a measure of condescension in the Chet Huntley school of intellectuals. Had they not squared themselves with the realistic demands of an environment terribly unforgiving of the weak, they would not likely have survived.

viii

Concurrently with the stomach cramps of Apollo's gestation period in the middle Sixties, as if these were not

*Petrone was an Army officer, but of the same breed.

enough, NASA also had had its hands more than full conducting the series of ten arduous Gemini flights, from March 1963 through November 1966. During the preceding Mercury one-man program from May 1961 through May 1963, much had been learned of immediate relevance to Gemini and finally Apollo.

Mercury had answered the larger questions of whether or not man could survive and function adequately in a weightless vacuum, survive reentry, and be recovered safely from the ocean.

Gemini demonstrated that two-man crews could carry out rendezvous and docking of two vehicles in Space, indispensable requirements for a Moon landing, change orbits and remain weightless on missions as long as two weeks without deleterious side effects, more than adequate for an eight-day Apollo voyage. And a youngster named Neil A. Armstrong had gotten a feel of Space flight in seven revolutions of the Earth on March 16, 1966, and a precious backlog of experience had been logged by 25 other astronauts on a variety of missions.

All of the Mercury and Gemini flights could be termed successful, in varying degrees, but in nearly all there had been moments of grave concern for crews and ground controllers. There were malfunctions, emergencies, and close calls, in which unmanned spacecraft would surely have been lost.

Three lessons were obvious. The human pilot in the "loop" was an asset, not the liability deplored by proponents of unmanned spaceflight. Secondly, Apollo hardware would have to be of a higher reliability than in Mercury and Gemini, where astronauts had been brushed by the wings of death on Space flights that were crudely elementary and less dangerous by comparison with Apollo's intricacy. Thirdly, the experience thus far gained provided no assurance of a successful landing on the Moon—only a reasonable possibility, a finely calculated risk, stopping just short of the reckless (in the opinion of the Soviets, decidedly on the reckless side).

Against this background, and an ebbing tide of the initial popular and Congressional enthusiasm which greeted the program, NASA was settling down into its stride for the final pull. Enough disillusionment with NASA's performance had set in to encourage the publication of books like *The Rise and Fall of the Space Age,* by Edwin Diamond, and to stir up restlessness among the natives on the Hill who were bankrolling Webb. But the confusion, ineptitude, waste, and duplication which had plagued the early management and organization of so vast an endeavor were giving way to tighter controls of the operation under experts gradually being fitted into the right niches, and through better "interface," wrought of hard-won experience, with industry. But it could not be said with confidence that Apollo's builders were out of the woods, moving on course, on time, toward an optimistic launch date of some time in 1968.

We know now that such a goal was illusory. If "bulled" through, it might well have been fatal. God was to work in "a mysterious way, his wonders to perform."

J. Leland "Lee" Atwood, North American Rockwell Corp.

Sam Hoffman, Rocketdyne

James Webb, Administrator, NASA, 1961–1969

Dr. Maxime A. Faget, NASA

Dr. Robert R. Gilruth, NASA

Dr. Charles Stark Draper, "Mr. Gyro," of M.I.T.

Dr. Wernher Von Braun, NASA, Father of the Big Boosters

Joseph G. Gavin, Jr., Grumman Aircraft Corp.

Upper left, Dr. John C. Houbolt, NASA, Winner of the Battle of the "Bug"; upper right, H. A. Storms, North American Rockwell Corp; left, Dr. Joseph F. Shea, NASA; lower left, George M. Low, NASA; lower right, William B. Bergen, North American Rockwell Corp.

Upper left, Dr. George E. Mueller, NASA, Associate Administrator, Manned Space Flight; upper right, Paul J. Castenholz, "Mr. J-2," Rocketdyne; right, Dr. Kurt H. Debus, NASA, Director, Kennedy Space Center; lower left, Eugene F. Kranz, NASA, Flight Director; lower right, Willard F. Rockwell, Jr., Chairman, North American Rockwell Corp.

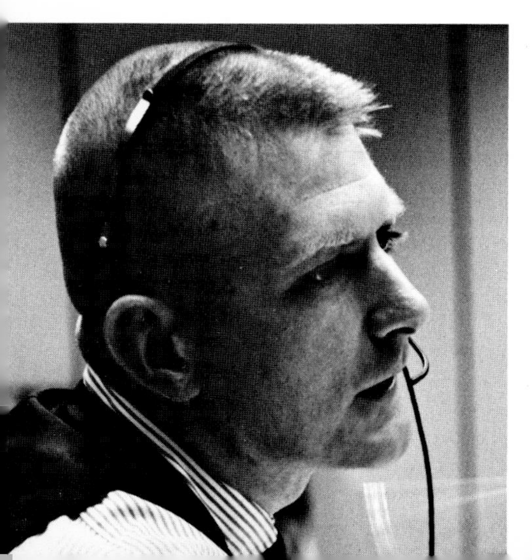

General Robert E. Greer, "Mr. S-II Stage," North American Rockwell Corp.

Lt. General Sam C. Phillips, USAF, Director, Apollo Program

APOLLO FLAG AT HALF-MAST (1967-1968)

"These three astronauts themselves . . . often said that the exploration of Space is a challenge worth the risk of life."

Dr. Wernher Von Braun

During the most painful months in its corporate history, following the first fatalities in America's Space program on Pad 34 on January 27, 1967, the leaders and craftsmen at North American Rockwell reacted much as might a boxer who has been floored by a hard punch in the middle rounds. Stunned, the company regained its feet and held on in a clinch. But ringsiders wondered whether or not it could go on to win the fight.

The dust having long since settled, no purpose would be served here in reopening old wounds or reviving old debates, merely for the record. But the reverberations from the more recent near-tragedy of Apollo 13 suggest a calm look backward at the manner in which the public, Congress, and the men of NASA and the prime contractor comported themselves in the wake of the far worse setback to Apollo, three years ago at this writing.

Both inflammable episodes refueled the tanks of the advocates of unmanned missions, for the advancement of scientific knowledge exclusively, or of the rechanneling of Space appropriations, forthwith, to more terrestrial goals. The voices would be heard afresh in the land of a Dr. Vannevar Bush, protesting as he had done in 1963 in a letter to *The New York Times:*

"To put a man on the Moon is folly, engendered by childish enthusiasm."

Coming when and especially *where* it did, on the ground, and coldblooded though it may sound to say so, the fire on the pad was providentially timed for the ultimate success of Apollo. The sacrifice of Virgil I. Grissom, Edward H. White, and Roger B. Chaffee deserves to be indelibly recalled as having probably spelled the difference in man's conquest of the Moon, safely, and in the saving of many more lives in Space than the three that were lost.

The deaths of five other astronauts from airplane and auto accidents, a tragic waste, have been nearly forgotten because they did not take place in a spacecraft. They were no less costly, however, to their families, to NASA, and to the nation.

Though high, the price paid on Pad 34 was far lower, quantitatively, than that which a single infantry patrol could expect to pay in lives during an ambush on any given day in Vietnam.

Popular support of a continuing, long-range program for the United States in Space was bound up in the answers to two questions about the 1967 fire, as it would be also, in a less urgent context, after Apollo 13's abort.

What happened? Why did it happen?

ii

What happened on Pad 34 seemed at the time both avoidable and unforgivable, as revealed by NASA's own investigation board and House and Senate committees on Space matters. There was no whitewash. In fact, newspaper editorials wondered aloud how NASA's own investigation could have been more damning in its published findings. Why the accident happened, in the perspective of today, is another matter. First, briefly, the "what."

Three astronauts died from asphyxiation (not burns) in a test of the spacecraft's compatibility with ground support equipment. Fire broke out in the cabin in a pressurized environment of 100 percent oxygen at 16.7 pounds per square inch, sealing the hatch too tightly to be opened after the fire had increased the pressure drastically.

The specific cause, as distinguished from the *source* of the fire, has never been determined. The source was attributed to an electrical arc or short circuit in the cabin wiring. Whether or not the wiring had been chaffed accidentally was conjectural.

No rescue or medical crews were on hand, since the test was considered nonhazardous.

The inward-opening hatch was inadequate for prompt escape of the crew. The cabin contained combustible materials. Both the wiring and coolant plumbing were vulnerable to fire or inadvertent damage.

The "why" of the accident, subject of prolonged and heated inquisitions on the Hill and in the news media, might easily have degenerated into a finger-pointing contest between NASA and North American Rockwell. In fact, Atwood received, but wisely rejected, proposals that the company mount a vigorous publicity campaign in its own defense. NASA likewise refrained from passing the buck. Both accepted the blame and closed ranks in an intensified effort to redesign the Apollo spacecraft, to improve safety procedures, and tighten management controls still further.

The contractor, for its part, could have reminded the public that it had originally recommended a two-gas atmosphere in the cabin and a quick-opening hatch, but had been overruled by NASA. The adoption of either of these options, certainly both of them, could have prevented the tragedy. And, as previously pointed out, the company had incorporated improvements in the wiring of its Block II version which would very probably have spared it the fiery fate of the cruder Block I spacecraft. The contractor could also have emphasized to the public that NASA had final responsibility for test, inspection, and acceptance of all hardware, and for the conditions that governed tests at the Cape.

In turn, NASA, while shouldering most of the blame, as it did for "blind spots" in its own vision, had extenuating circumstances in its favor. Chief among these, and the underlying cause worth remembering, is that *everybody* most concerned—NASA, the contractor, and the astronauts, who had been given every opportunity to express themselves on safety measures—believed a fire hazard had been eliminated. Frank Borman, of NASA's crew safety committee, told Congress:

"We tried to identify every hazard we could find, but this one we missed."

NASA's blind spot was an understandable reflection, not of ignorance, but of its satisfactory experience with 100 percent oxygen in the environments of Mercury and Gemini. There were no fires, in countless tests and hundreds of hours of Space flight.* Any fire requires, first, a source of ignition; with proper insulation, no electrical wiring could foreseeably cause a fire. The insulation was deemed adequate, ergo, don't worry too much about inflammable materials in a cabin that won't catch fire anyway—trade off on weight savings.

The safety merits of inflammable oxygen over a two-gas atmosphere were, and still are, a matter of valid debate, of prudent, closely reasoned trade offs. One is safer on the ground. The other in Space.

The same kind of trade off holds true of two hatch designs—inward and outward. Grissom nearly drowned because the outward hatch of his Mercury capsule, designed for quick exit, inadvertently exploded open after splashdown. The near-tragedy bore heavily on NASA's decison to instruct North American to revise its

*913 hours, 20 minutes.

design to an inward-opening hatch for Apollo. Consequently, Grissom got the worst of both bargains, in one of history's grimmer ironies. For each emergency, he had the wrong hatch at the wrong time. The conclusion is reasonable that there was no ideal choice for NASA in the foregoing dilemmas. Apollo's atmosphere is still composed of fire-hazardous 100 percent oxygen, once the astronauts have spent the first few minutes of launch in a two-gas atmosphere—a further refinement in the trade off game.

iii

Both Webb and Atwood drew heavy fire for arguing before Senator Anderson's committee about the existence of a "Phillips Report." Both were technically correct in referring to the findings of Phillips' survey of 1965-1966 as "Notes"—the only term used throughout the Phillips letter to Atwood, which was quoted in full in the preceding chapter. And there is something to be said in rebuttal of those, and there were many, who impugned the motives of Webb and Atwood as purely self-serving, in the harshest light.

No leader of a large organization would be worth the TNT to blow him to hell if he lacked the loyalty to try to protect his team's good name, under fire, by fighting against the public disclosure of damaging evidence of internal problems—*provided* that remedial actions were going to be initiated every bit as rigorously without such disclosure as because of them. The death of three astronauts had provided Webb and Atwood with a far more powerful incentive to root out any chance of a recurrence than any censure could bring.

Both men knew this. Both also knew that Atwood's response to Phillips' complaints back in 1965 had resulted after only four months in a turnaround, with Phillips reporting encouraging improvement in the company's performance. Both men displayed the courage of their convictions, in a dark hour, that their respective organizations could eventually discharge their responsibilities, as they have since magnificently demonstrated. I give them credit for sincerely believing that, in their share of so great a national commitment, only a destructive purpose would have been served by exhuming a two-year old resume of difficulties experienced during the growing pains of Apollo. The final report of Senator Anderson's committee published in 1968 supports such an interpretation in these categorical words:

That according to its NASA witnesses, the findings of the Phillips task force had "had no effect on the accident, did not lead to the accident, and *were not related to the accident* [italics mine]."

Another general observation needs to be put forward. If anything is absolutely predictable, it is that no matter what the endeavor, men will make mistakes. Pete Everest, veteran test pilot of the Bell X-1A rocket plane at Edwards Air Force Base, was not known for making glaring mistakes; yet he admitted a case of 100 percent pilot error when he landed his F-100 fighter one day at Edwards with the gear up. A brilliant eye surgeon ruined his career when he removed the *wrong eye* from a patient.

I feel more inclined, in perspective, to ask, not "who were the culprits responsible for the fatalities which occurred on Pad 34?" but "what kind of culprits, from the same cast of characters, have been responsible for *no* fatalities among the occupants of a spacecraft with six million functioning parts that could have failed, in the manned missions of Apollos 7, 8, 9, 10, 11, 12, and 13?"

In the book *Why We Went to the Moon,* written by three journalists of the London *Times,* in which the gloomiest construction is put upon the motivations of nearly all of the leading figures in the Apollo program, from President Kennedy on down, the statement appears that according to NASA's accident-investigation philosophy, "*People* were not responsible. Machinery had merely failed to work."

Their point strikes me as spurious on two counts. First, nobody knows better than NASA that machinery is not "responsible" for itself. Second, two high level people were fired, by whatever name one chooses to call it. Or have the authors not heard of the prompt replacement of Joseph F. Shea, the man who had been directly responsible for Apollo's command module at the Manned Spacecraft Center, Houston, by George Low, a NASA veteran from the Eisenhower regime?

Shea's opposite number on the industry side likewise became a casualty—Stormy Storms.

The book also pointedly raises the question of why Shea was not called before Congress to testify, speculating archly on the reason. What the authors failed to say is that Shea was suffering a virtual, if not clinical, nervous breakdown after a devastating blow of a kind from which even the strongest individual is not likely ever to fully recover.

Friends of Shea and Storms have said that these two became the inevitable scapegoats of the 1967 fire because they were the most proximate to the line of fire. Less friendly commentators might say that they probably had it coming to them. To me, they were both as much human sacrifices as the three astronauts to the reverses to which man exposes himself, voluntarily, when he tests his capacities to the ultimate. On the ground in the laboratory, the factory, or at Mission Control. Or out in Space.

Drastic measures were in order, and were speedily forthcoming, to crack down hard on carelessness or faulty procedures where they had been uncovered in the Apollo-204 investigation. But in a broader sense, I can only take a dim view of the more vociferous among the Monday morning quarterbacks in 1967, who never in their lives had confronted complexities to compare with those which harassed the builders of Apollo, yet who threw up their hands in horror at the presumed shortcomings and motives of better men than they.

iv

In the emotional backwash during the months following the fatal fire, from Atwood's executive floor, to

Storms office at Downey, to the cavernous buildings where the Apollo Command Module was being redesigned, managers, engineers, and lowest paid craftsmen alike shared a kind of shell shock.

Hurt pride, shaken confidence in oneself and in one's superiors, and difficulty in regaining a sense of direction, of reality, after what had happened, gradually gave way to a spirit of "we'll show 'em," and a deep new urge among the people of NASA and the contractor to pull together with a single eye to victory. There was nothing to be gained by, nor was there time for, recriminations. "You think you guys were putting out before?" one grizzled supervisor growled at his team. "Well, hold on for the ride, we're going to get in shape."

A new broom swept the buildings at Downey, not clean of key employees, most of whom were retained by William Bergen, Storms' replacement, but clean of whatever cobwebs or fat Bergen could find.

Atwood has always had a reputation for agonizing reluctance, and sometimes inability, to force himself to replace a veteran executive. This was never truer or more excruciating to him than in the case of Storms, whom Atwood believed had labored long, well, and above the call of duty. Webb, on the other hand, can hardly be faulted for insisting that, at the very least, there be a changing of the guard in the president's office at Space Division. Still, Atwood resisted stoutly to the end, bowing only to the very real prospect that North American's contract might be transferred to another contractor, with all of the grave implications, not only for his own company but for NASA, attendant on perhaps an additional year's loss of priceless time in a schedule that would now have to be slipped from one to two years in any event.

Bergen talked to me freely about the circumstances surrounding the change of command, when I visited him at Downey a year ago.

Before joining North American as a corporate vice-president in April 1967, William B. Bergen had been that rarity, an ex-president of a large corporation out job hunting. Voluntarily.

Bergen's departure from the Martin Company in 1966, where his record as president had been distinguished, was occasioned by certain "policy" differences with chairman George M. Bunker, which can and do arise between two strong executives, without discredit to either. An old hand with airplanes, he was looking forward to finding a spot with some outfit like McDonnell Aircraft—where he could indulge his professional "hobby" and enjoy life "before I got too old," as he put it (he was then in his fifties)—when a phone call from Atwood tossed him right back into the pressure cooker.

As the recent president of a major aerospace firm, Bergen happened to have exactly the proper credentials to fill the newly created post at North American of corporate group vice-president, with the presidents of Rocketdyne and Space Divisions to be working under his supervision. He had barely warmed his chair in that prestigious capacity when he received an urgent call from Atwood, a month later.

"I've got no choice left, Bill," he said painfully, "but to put somebody else in Stormy's job. What am I going to do?"

"You're going to *demote* me," Bergen answered without hesitation, "and I'll go down there and run Space Division for you."

There was a long pause.

"Bless you, Bill," Atwood said in a low voice.

For many months thereafter, the office in which we were talking had served as working, living, and sleeping quarters for Bill Bergen, so that he could go out into the plant, if necessary at three o'clock in the morning, to render prompt decisions.

First priority was recovery from the staggering blow to morale that pervaded the plant. He moved in swiftly to restore confidence in management and pride, and, as in the case of Greer before him at Seal Beach, to install a less hectic, more businesslike administration. Where necessary, he trimmed out some of the "old guard" types who had been in their positions too long, blocking the advancement of more eager, younger types.

None of this should be construed as disparaging to his predecessor, whom Bergen likes and admires. He was keenly aware of the double-whammy that had pounded on Storms—a tremendous expansion of Space Division after 1961, a "people explosion" on a nearly crash basis to cope with two huge contracts at once, Apollo and Saturn II, compounded by the task of design and development of a completely new kind of maddeningly delicate hardware.

Storms has told me himself that the time can come when a change of leadership teams is good for everybody. He had done his damndest and that was that. No bitterness. Bergen agreed with those I've heard appraise Storms as probably tops when it comes to breaking the back of difficult new programs, but less at home once the smoke has cleared and a man of different temperament is needed for the stage of more normal, day-to-day operations. Bergen was that man.

My notes thumbnail him as: "Fit. Clear-eyed. Face unlined. Lunch consists of running a mile instead of eating. Calm. Deliberate. Accurate. Fierce concentration. Unlikely to get caught with pants down under any circumstances. Sharp judge of human nature. Spots phonies in real time. Quiet sense of humor."

While continuing to elicit standout performances from wheelhorses of the previous administration like Myers, Ryker, and Feltz, he transfused some new blood into the division with first Ralph Ruud and later Joe McNamara as his number two, George Jeffs as chief engineer for the Apollo CSM, and John Healey for Manufacturing.* A new head of Quality and Reliability Assurance, T. C.

*Ralph Ruud continued to provide close high-level coordination.

McDermott, replaced test pilot "Scotty" Crossfield of X-15 fame, who had seemed to be ideally qualified for that assignment. Crossfield's removal—given the prevailing atmosphere, after the fire, of near-hysteria throughout the Apollo program on the subject of reliability—was practically foreordained.

"When the eggs hit the fan," I was told by John Young, the vice-president of North American most intimately concerned with reliability, who nonetheless managed to weather the storm, "the quality control man is traditionally the first to get fired. Everybody tromps on him. After all the years of flat-out emphasis on reliability approaching a frenzied overemphasis at times, our guys were demoralized by the barrage of criticism. Maybe Scotty was overwhelmed by large events—he's a good man. In any case, I was told to get hold of the best quality man in the United States, to go outside the company if necessary. The eventual choice was McDermott, from our own Autonetics Division."

After leaving Bergen, I spent some rewarding time with John Healey, who is about as far as you can get from the stereotype of a stolid engineer. He is a veritable geyser of energy, ideas, eloquence, and extroverted enthusiasm, who could have been a skilled performer in the less businesslike arts.

Healey's first assignment, after fellow Martin alumnus Bergen brought him in from Martin-Marietta, was that of bird dog. He was to spend day and night, if need be, and he frequently did, shepherding the Apollo-101 spacecraft through its assembly to the final combined systems checkout in Building 290. The redesigned bird entrusted to his care was scheduled to sit atop the Stack for the first manned flight in 1968. He wasted no time in making his presence known, armed with *carte blanche* from Bergen to make on-the-spot decisions and to set the tone in his capacity as assistant Apollo spacecraft program manager.

Healey's critical eye came to rest on a white rope cordoning off against visitors the gleaming, silver Apollo cabin under construction in a sanitized, sacrosanct area of Building 290. A lady employee in a white smock, with a book in front of her, guarded the roped-off area.

"What's the book for?" Healey asked her.

"It's a register, for visitors," she said.

"*What* visitors?"

"Why—uh—they're generally VIP's."

"Sort of like a hotel register, or a guest book at a wedding?"

"Why, if you want to call it that, it's for the record."

"Listen carefully," Healey said."It's three o'clock. I'll be back at four o'clock. I want to find the rope and that guest book gone."

"But I can't do that," she remonstrated. "Not without permission from my boss."

"I'm your boss, everybody's boss around this area. The name's Healey. Spread the word around. There'll be no one allowed in that spacecraft except the six men working on it. No visitors. Not even General Phillips or Mr. Bergen, without my personal approval."

He was back at four o'clock. The rope and book were gone.

The word got around, following which the only VIP's given access to the spacecraft's interior were the six craftsmen on each of the eight-hour shifts. They were to suffer no interruptions, even to go personally after a tool—if need be, some vice-president could fetch it for them.

"Every tool was accounted for when a man checked in," Healey told me. "If a tiny screwdriver was missing when he checked out, nobody went home until it was found."

Charlie Feltz had mentioned to me that adding even one apple apiece to the astronauts' menu, as had been suggested, could be a serious matter for the engineers. I asked Healey to elaborate.

"Well," he said, "I can rattle off several ramifications for you. One, you process a change request through NASA. Two, you consult the experts in weight analysis—we deal in precious ounces. Those apples represent extra pounds you'd be toting around for half a million miles on a Moon trip. So, three, there has to be an analysis to determine if even that small amount of extra weight involves additional fuel. Four, you may have to redesign your food containers to accommodate the apples and keep them from floating away and lodging where they could cause trouble. Five, the materials people must be consulted. Six, the question of the stowage installation—how, what attachments, where mounted? Seven, all this must be integrated with the existing configuration of the spacecraft, for possible interference with some other system. Eight, if you've altered the food stowage installation, you'll need new sets of drawings and there's a new item on the checklist. Okay, there's eight ramifications for you."

"Aren't you exaggerating a bit?" I asked him.

"A little, but not much," he said. "I'm just trying to give you some idea of how one small change can entail dozens of other changes, or how dropping a major change into the pond can send ripples all the way to the beach."

Healey made a point of going to each subordinate, including the workman on the job, and asking, "What do you need?" If the man was waiting on a part scheduled for delivery in a week, he'd tell him, "I'll get it for you tonight," pick up the phone and have it flown in from Michigan if necessary.

He felt strongly that what someone else tells you or what you read, no matter how authoritative, can be a poor substitute for personal knowledge, for "getting your hands dirty," and he liked to ask:

"Have you put your *hand* on it, touched it?"

Knowing that he *was* the company, not a pal, he gave terse answers to anyone close to him that tried to take advantage.

"I'm a couple of days behind schedule," one of these informed him, "but don't worry, John. I can catch up. Okay?"

"*Not* okay."

Thus, under the spur of Bergen's team, imbued with the same elementary, but frequently neglected management philosophies as those of Healey, Space Division quickened its pace toward the day when Launch Director Rocco Petrone could affirm: "All systems Go."

That day, it was now clear, had slipped by approximately a year and a half. But for North American Rockwell, Grumman, M.I.T., and many other builders, including NASA of course, that time would be well spent.

V

In the waning months of 1967 the flag over North American rose slowly, figuratively, from half-mast. All was not black. Saturn II was coming into the groove. Sam Hoffman's F-1, H-1, and J-2 engines, after thousands of test firings each, had survived the hawkeyed scrutiny of Von Braun, Eberhard Rees, Oswald Lange, and General Ed O'Connor at Huntsville, and of "Chuck" Allen's static test conductors at the towering stands of the Mississippi Test Facility where barges were towed up from New Orleans. Then the engines had fired successfully in unmanned launches of Saturn I and the Saturn V at Cape Kennedy, in man-rating tests for Apollo. And at Houston, Rocketdyne's smaller but equally critical "thruster" engines, which can fire several hundred times a second in controlling the Command and Service Modules, were passing muster with George Low.*

From another quarter, Atwood likewise received a welcome shot in the arm, a vote of confidence at the psychological moment, from Willard F. Rockwell, venerable chairman of Rockwell-Standard, Inc., known as the "Colonel," and his son "Al" Rockwell, Jr., president. Their firm had been negotiating a merger with North American.

*Marquardt supplied the thrusters for the Service Module, under subcontract to North American.

Al Rockwell told me of pressures on him and his father, after the fire, to reconsider the merger. Certain stockholders didn't think that Rockwell-Standard should identify itself with a "loser" whose image had presumably been tarnished by bad publicity. Furthermore, the Justice Department was throwing a roadblock into the path of the merger by insisting that Rockwell-Standard divest itself of its executive-type Jet Commander, which Justice deemed to be in competition with North American's Sabreliner, albeit a heavier and more expensive aircraft. This position the good Colonel stoutly protested, pointing out also that a severe and arbitrary loss would be entailed in a forced sale of its Jet Commander operation at Bethany, Oklahoma. But in vain.

In spite of these deterrents, the Rockwells stuck by their guns and consummated the union, thereby serving notice on all, not least Atwood's employees, that responsible analysts believed North American would ride out the storm and once more uphold its long reputation for excellence.

What, if anything, did this contribute directly to the success of Apollo so late in the game? In terms of hardware, nothing—no truck axles were needed. But in terms of such intangibles as confidence in the future for a company whose powerful prestige had been beleagured, and in some measure in its effect on the company's remaining performance on Apollo, perhaps a great deal.

New blood in management from the Rockwell side, of that chillier temperature which is salutary in difficult times, was provided by Robert Anderson, dispatched from Pittsburgh to El Segundo as executive vice-president. A measure of Anderson's versatility and cool-headed capabilities is reflected in a fast track record. The Rockwells recruited him from top management at Chrysler Corporation in 1967 and sent him to North American in 1968, where he rose swiftly from executive vice-president to president in 1970, after Lee Atwood reached retirement age.

vi

Even though it leaves a question mark, this chapter ought not to conclude without relating the visit I paid to Joseph F. Shea, for part of him died on Pad 34 with the three astronauts who had depended on him, more than any other, for a safe spacecraft.

Could he, or would he, shed light on any dark corners that in the absence of his testimony before Congress had been overlooked, I wondered, as I drove from Cambridge, Mass., along the Charles River in a pouring rain on my way to a 10:00 P.M. interview at Joe Shea's house. He lived 20 miles out in the country near the Raytheon plant in Waltham, where, after leaving NASA, he had joined Dr. Brainerd Holmes as a vice-president at Raytheon, builder of vital computers and other electronics for the Command and Service Modules.

It was a miserable night, with oncoming headlights glaring on the rain-smeared windshield, contributing to my gloomy foreboding that I might be embarked on a fool's errand. What if he proved embittered and distrustful of a strange writer, unwilling to divulge anything of importance? Was I unduly imposing on him in having agreed to an appointment at so late an hour, when presumably he'd be plumb tuckered out.

One thing was encouraging. He was clearly not the kind of man who set much store by his own personal convenience. He'd set the awkward hour because it was the only free time he could squeeze in. He'd just returned that morning from a business trip, was delivering an after-dinner speech at the moment at a nearby college, and would be leaving first thing the next morning on another trip. Yet for some reason, he'd been willing to forego his privacy and a brief respite at home. Maybe it would be worth the discomfiture, after all, of having passed up the customary cocktail hour, out of deference to my upcoming interview.

Arriving a little before ten, and observing no car in the

Shea's garage area, I parked under a dripping elm tree and waited until his car appeared and he had gone inside, just as my car radio pronounced that it was exactly ten o'clock. Answering the door and greeting me courteously was a tall, athletic man with the look of command and the stamp of keen intelligence in his gaze, also with the three-cornered fighter's eyes I'd remarked in Max Faget. He introduced me in the living room to his gentle-eyed wife, a young-looking mother of five children. Of the two, she appeared to be the wearier, although he had been through an extra-taxing two days, and there was a wistfulness in her smile when she excused herself. In a flash, I felt that her expressive face betokened a story in itself, of tribulations that men are fortunate not to have had to bear alone.

"Care for a drink?" he asked expectantly.

"That I do," I said. "I'm dry as a sponge. I didn't want to drink before an interview."

"That makes two of us," he said, "on account of my speech tonight."

"How'd it go?" I asked.

"Always hard to tell," he said, "but nobody walked out."

He excused himself and reappeared shortly in shirt-sleeves, bearing a pitcher of martinis and two tumbler-sized glasses. Guarded at first, and a reserved man in any case, he gradually warmed up to the topic uppermost on our minds—what it had taken and what it would take for men to land on the Moon. It was clear how strongly there still clung to him a sense of responsibility from those critical years when he had been known at Houston as Mr. Apollo—for 24 hours a day nothing else had existed for him but his colossal obligation, only in the end to taste ashes.

He said nothing critical of North American or Storms, as Storms had refrained from criticizing Shea to me, although I knew that there had been no holds barred in many a spirited difference of opinion between contractor and NASA. When I alluded to this, he nodded and said:

"You should see some of the transcriptions of our communications."

He was, however, hypercritical of the human weakness of impulsiveness or the haste that can make waste in a Space program.

"You eliminate the unknowns," he said, "or you don't go. I'm talking about the hardware unknowns that *can* be eliminated, of course. Repeating a mistake that's been made before because you haven't really spelled out the problem is unforgivable. Decisions short of the ideal have to be made, or you may never get the job done, but you must *know*, not *assume* anything that's capable of being tested out."

"Like fire hazards?"

"Yes, the early cabin design should have incorporated exhaustive ground tests that were not made."

Referring to the book I envisioned, he said frankly:

"I doubt that you'll be able to get the whole story about the builders of Apollo, because they won't tell you the truth."

"I don't think anybody's lied to me yet," I said.

"I'm sure they haven't," he said. "I mean omission."

I waited for him to go on. But apparently he meant to include himself in their number, unwilling to point a finger at others. But the message was clear. By Shea's standards, the Apollo job could and should have been done better from the start. Among his peers, I told myself, he was entitled to that opinion. But in the less Olympian, less rarefied atmosphere of the layman the standards already stood formidably high. That he thought they could and should be higher gave this interviewer fresh grounds for optimism that his successors would win out. I felt honored to be sitting across the coffee table from a gladiator who, though wounded in the arena, still symbolized that hope.

Two hours later, after discussion of more technical

matters like "mathematical models" for computer analyses of complex systems, he led me to his den and pointed to a spacecraft model.

"Don't fail to give Grumman its due," he said, touching the lunar module. "Few people have any conception of what that LM represents, a whole separate spacecraft in itself. Terrific achievement."

His words were to come back to me during the "lifeboat" rescue of Apollo 13 in 1970. Then we stopped before a color photograph of Grissom, White, and Chaffee in their white spacesuits, heads bowed in reverent prayer at a table flanked by an American flag, taken at Downey on the occasion of NASA's formal acceptance of Spacecraft 12, five months before the fire. It was autographed humorously to the effect that, "It isn't that we don't trust you, Joe, but this time we've decided to go over your head."

When he spoke, his eyes and voice were charged with emotion.

"How's that for *irony*?" he asked. Then he cleared his throat and added:

"It didn't have to happen."

Driving back to the motel, pondering his parting words, I believed that I knew what he meant—regardless of where the blame lay, or of how many men shared it, or of the permanent contribution for which many associates have given him grateful credit, he would never forgive *himself.*

FROM THE J-2 MYSTERY TO APOLLO 8 (1968-1969)

"Eleven more licks'll fetch 'er.

Uncle Remus

The year 1968 began and ended on a note of risk-taking that makes one shudder in the light of what was to happen two years later to Apollo 13. If bold, calculated risks had not been accepted by NASA, however, and had it not been for a remarkable bit of scientific detective work to unravel the baffling cause of failures of the J-2 liquid hydrogen engine in Space, any hope of meeting Kennedy's "decade" deadline might easily have evaporated.

Was the making up of lost time all that important, in 1968, to meet a target date of 1969 if humanly possible? Critics of haste still say no, even if the Soviets had been in the race, which they were not.

Jim Webb and most NASA managers would argue that the jockey has got to know where the finish line is, that men work better against demanding deadlines leading to a firm and highly visible goal, and that indeed a schedule for a large-scale endeavor that lacks urgency is an invitation for all hands to rest on their oars—essentially no plan at all.

Certainly the principle would appear to have been sound, psychologically, that men do not reach peak performance on a business-as-usual, nine-to-five routine—that there must be a clear, sharp focus to elicit that extra effort needed to match the task, a demanding timetable that *compels* it. NASA's top echelon was responding to that compulsion.

More painfully than any other, as the highest-ranking decision maker concerned solely with the Apollo program, Sam Phillips weighed the scales of risk, attentive to voices from both subordinates and superiors. There were two ways that he could speed things up appreciably. He could reduce the number of unmanned test flights of Apollo from the seven originally contemplated to as few as two, if both were winners. Secondly, instead of waiting for Grumman to iron out all remaining bugs in the "Bug," he could send off the initial manned flights with an abbreviated Stack, minus the Lunar Module. The latter, designed specifically for the lunar landing task rather than as a "lifeboat," was then thought of as more of an appendage than a potential substitute for the Command Module, with the latter's far more elaborate systems. Few, if any, planners anticipated how dramatically the tail would one day be called upon to wag the dog. Providentially, however, they had insisted that the LM duplicate the Command Module's navigational and life support systems on a "minicar" scale.

Dr. George Mueller, Phillips' immediate boss, and Deputy Administrator Thomas O. Paine (subsequently administrator) were to back up Phillips in their recommendations to Webb on these tempting but nerve-jangling options.

In the optimistic aftermath of what was hailed as a successful maiden flight of the unmanned Apollo-Saturn AS-501 Stack from Pad 39 during the previous November, General Phillips concluded that only one, instead of six more unmanned flights of Apollo-Saturn would suffice to man-rate the Stack—a drastic cut that would make possible a great leap forward in the schedule. If that second test, scheduled for April 4, 1968, were equally trouble-free, firm plans could then be made for Wally Schirra, Donn Eisele, and Walter Cunningham to ride atop a Saturn IB booster on the first manned, Earth orbital mission, before year's end—without waiting for the LM.*

Unfortunately, the "partially successful" flight of the second Apollo-Saturn, AS-502, in April 1968, turned out to be a bewildering disappointment in one crucial aspect. Failures occurred during the operation of a J-2 engine in both the second and third stages of the Saturn V. The bare facts of what happened were summarized by builder Rocketdyne as follows:

"Following successful first stage S-1C operation, the cluster of five J-2 engines in the S-II stage started properly and achieved correct mainstage operation. After approximately 70 seconds of normal operation, chilling of the engine compartment in the area of outboard engine No. 2 began. This chilling was followed by a slight loss in engine performance as noted by reduced chamber pressure. Then, 193 seconds later, the No. 2 engine lost pressure, causing it to signal premature shutdown of this engine. Shutdown was achieved safely. Due to a stage wiring condition, the shutdown of engine No. 2 caused engine No. 3, which was operating normally, to also shut down prematurely. The remaining three engines were not affected by any abnormal effects of engine No. 2 and continued to operate properly for the remainder of the required burn time.

"The S-IVB stage single J-2 engine start was successful, followed by normal operation for approximately 68 seconds. At this time, abnormal engine compartment chilling occurred, followed 40 seconds later by a slight chamber pressure performance decay. Engine operation

*The two-stage Saturn IB is, of course, smaller than the full three-stage Saturn V, and had logged more flights by far, using a single J-2 engine in the S-IVB second stage.

then continued for the intended duration of the first burn (total of 170 seconds), followed by satisfactory engine shutdown. Following normal orbital coast, the engine failed to achieve proper ignition upon signal to restart and the sequence terminated."

Most encouraging of the "partial" successes of the mission was the performance of IBM's IU, or Instrument Unit, the Stack's brain, in adjusting to the emergency. The brain issued commands that kept the rocket on course during the climb-out and enabled it to reach Earth orbit despite the loss of power and the uneven thrust, making possible the continuance of the mission.

Rocketdyne's summary, written by Paul D. Castenholz, J-2 program manager, went on to say of the investigation and testing which followed:

". . . Duplication of the flight failures *could not be achieved* [italics mine] during extensive testing at ground test conditions. Testing at these conditions masked the apparent problem, and the environment 'protected' the fuel line."

How, then, *was* duplication of the failure achieved? How was the villain caught and exposed?

No doubt Sherlock Holmes and Watson would have been stunned with admiration for the ingenuity of the sleuths—Saverio Morea's J-2 team at Huntsville, working closely for NASA with Paul Castenholz's J-2 engine builders at Rocketdyne—with all of the hardware lost in Space, and only the meagerest of clues from the telemetry data that had been received back on Earth.

Since any prospect of manned flights in 1968 was now abruptly out of the question, pending a pinpoint identification of the failure and its correction, Morea and Castenholz found themselves roasting over an open flame ignited by Von Braun, who called for a round-the-clock, seven-day-a-week effort by the contractor and NASA to track down the villain. An intense and high-geared performer by temperament, Paul Castenholz in particular had to step up his naturally feverish pace to new heights, pushing himself and others almost beyond endurance.

ii

Lee Atwood placed every resource of North American Rockwell at the disposal of Sam Hoffman and Bill Guy over at Rocketdyne Division, who in turn gave Castenholz's troops top priority. What followed is a classic example of engineers unleashed in full cry after their favorite quarry—a fact.

Now a full-dress technological witchhunt was mounted, beginning with "rounding up all of the usual suspects." The first of these was "quality." Had there been defective materials or workmanship anywhere in the J-2's innards, and especially on stainless steel tubing in the vicinity of the ASI (Augmented Sparkplug Igniter) fuel line? Here was where the drop in temperature had been recorded, presumably from a leak of the supercold liquid hydrogen, or an outright rupture of the line. Not much to go on, but a starting point.

The sleuths went all the way back to their smallest suppliers for certification of tubing material, bellows material, and manufacture. Over 300 fuel lines were analyzed, and X rays of the welds and lines used in AS-502 were reviewed, as well as other lines fabricated in the same time period. The analysts came up empty. No anomalies. No clues.

On the assumption that the ASI fuel line assembly might have failed in flight at the extreme limits of the engine's range, 85 test firings were performed with extensive instrumentation for complete examination of all operating characteristics. Each assembly was tested in excess of 4,100 seconds. No failures. No clues.

They tried to simulate the failures by normal and then very abnormal flow rates of fuel through the ASI line. Only at approximately double the normal flow rate did they succeed in damaging the bellows section of the line—an accordionlike series of convolutions designed to accommodate stresses and make the line flexible. No help. The results did not simulate the flight results.

They vibrated the hell out of a series of lines at

frequencies and amplitudes well in excess of engine-induced levels. The same for an entire engine, simulating flight conditions. No damage.

They forced gaseous nitrogen to flow through the lines at excessive velocity. Again, no damage occurred except at flow rates higher than those encountered in flight. But they did find the first glimmering of a clue: the existence of a "resonance" condition was noted. As every schoolboy knows, from the practice of troops not marching over a bridge in step, every metal can fracture if it is resonated at exactly the right frequency, at a certain "harmonic."

It was then found that the ASI line had several operating points where the amplitude of vibration was higher than normal. But frustratingly, the line could not be made to fail even at the level of the worst vibration. Castenholz's official summary of that frustration:

"From the experiments and studies conducted thus far, no operating conditions could be found for the ASI fuel line that would simulate or explain the flight failure. Even though there was an indication that a resonant condition existed at the upper end of the flow operating region, there was apparently still a phenomenon which occurred during flight that had not been simulated in the ground testing accomplished to date."

The trail was still cold, despite the combined efforts of two teams, one for the S-II failure and one for the S-IVB, concurrently checking out the performance of the thrust chambers, turbopumps, valves, gas generators, and other components. The sleuths now pressed forward more singlemindedly on the most promising assumption—that the line had been severed or ruptured in some way that could be simulated to duplicate the flight data. Okay, so you start by cutting a line in two and observing what happens.

The fundamental approach of Castenholz and Morea now became that of trying to match up two sets of "fingerprints." They were in possession of the master fingerprint, namely a second-by-second record of the sequence of events in the two J-2 engine failures. Possibly, by inducing failures in actual firings of the engine, until, hopefully, one of them precisely duplicated the malfunction in Space, they would have their matching fingerprint. They could then be positive of what had happened and make such specific engineering changes as would rule out a recurrence. Nothing less, no educated guess, could release Apollo from the existing "hold" on manned missions.

The investigators were well aware that deliberately slashing a fuel line and then burning a 200,000-pound thrust J-2 engine was no child's play. An explosion could damage the test stand to the tune of millions of dollars, to say nothing of the hazard to test crews, and more than one engine might have to be destroyed at a cost of $1.5 million per copy (three, as it turned out, were tested, but not destroyed).

With an eye to the taxpayers' money and to the safety of the technicians, the engineers surrounded the "suspect" upper section of the propulsion system with protective shields, including boilerplate, to localize damage from fire or blast. In doing so, they took great pains, however, not to alter any conditions or sequences of operations in an important respect that might lead to false conclusions; a test that didn't duplicate the crime in all respects would be inconclusive. The tests began.

Cutting the ASI line cleanly caused a failure, all right. But the resulting fingerprint didn't properly match the master print. Failure came too quickly, and there were other discrepancies. Perhaps, the sleuths then reasoned, the line had ruptured gradually instead of sustaining a clean break.

Next they tried weakening the line structurally, so that it took longer to fail. The fingerprint came a little closer to matching. Still not close enough, however. So the line was modified to hold its strength a little longer still, before failure.

Bingo! Part of the fingerprint now matched. Here at last was something concrete to go on. They had the guilty fuel line pinpointed, and they knew that it had indeed ruptured gradually. But the question still remained: what had *caused* it to rupture gradually?

The analysts were now fresh out of just about every hypothesis but one. Something had happened out there in Space that had never been simulated. Did a hard vacuum, then, hold the key to the mystery? Or part of the key? The J-2 had been fired in a vacuum on the ground, but the tests had included only the combustion chamber, not the accessories like the upper assembly of the ASI line.

"Why not?" I asked Castenholz.

"Bear in mind first," he said, "that the J-2 had already passed a severer test. It had flown on missions in Space without giving any trouble. Which would imply that the environment of a vacuum was irrelevant. But if the vacuum environment in Space *did* have something to do with the failures, you were up against the question of why didn't you get a busted line every time? We figured there'd have to be at least one other factor, like some small difference between one engine and another coming off the production line. Second, no engineer in the world would have foreseen any conceivable need to test the ASI line in a vacuum—it's just an inert hunk of stainless steel. Engineers test only for contingencies they can conceive of."

Following up their latest lead, Rocketdyne's researchers built an elaborate test stand at its Santa Susana facility to observe the performance of the ASI line in a man-made vacuum that came as close as possible to simulating a hard Space vacuum. Eight lines were subjected to the tests. Within less than 100 seconds, and at the same flow rate of fuel as had induced the natural vibration frequency of the lines, *all eight* showed a failure in the bellows section.

This simulated closely the engine environment and obtained the failure results duplicating the failures on AS-502.

At last, the fingerprints matched. But instead of being pierced, the mystery was to become even deeper. If vibration was the culprit, why wasn't it *consistent?* Why did it occur destructively in a vacuum only? Why, at exactly the same frequency, at the same flow rate, were the engineers unable to precipitate a failure when the bellows section was surrounded by a normal Earth atmosphere? Why? Why? Why?

Something *had* to be present in the atmosphere that was protecting or insulating the bellows from vibrating. Could it be moisture in the air? Possible. Or liquid air forming on the convolutions of the bellows? Also possible. The temperature of the hydrogen fuel flowing through the bellows was more than adequate, minus 350 to minus 400 degrees Fahrenheit, to liquefy the air coming into contact with the stainless steel.

First, photographs were taken of the bellows while exposed to a normal atmosphere, with the fuel flowing. Sure enough, the bellows accumulated a sort of frozen "sludge," between the convolutions, which prevented vibration, like "padding." No failures occurred. This was nearly, but not wholly, conclusive evidence for the detectives. They still had to eliminate other possible factors, like moisture—or atmospheric pressure.

A test of high atmospheric pressure on the bellows, but with helium gas instead of air as the environment (helium does not liquefy at low temperature) did not protect the bellows. It failed rapidly. So a high atmospheric pressure had not been beclouding the issue in ground tests.

Next, Castenholz simulated an environment where only water vapor (and no air) was in the region around the bellows. Some ice formed, but not enough to "protect" the bellows, which ruptured in due course from vibration. So moisture alone had not been concealing the real villain.

Final and conclusive proof that liquefied air was the villain emerged from a test of the effects of "heat transfer" from the liquid air through the walls of the bellows to the liquid hydrogen at different flow rates. It too helped to throw the destructive vibrations off their stride, in addition to the damping or padding effect of the "sludge."

Q.E.D. The villain had come out of the shack with its hands up.

How long did all this take? Thirty days. It might have taken thirty weeks, or longer, had it not been for assiduous use of the computer. As Paul Castenholz explained it to me, an engineer can describe an engine mathematically, just as you or I would describe it in words, but more precisely. From the resulting "mathematical model," then, an almost infinite variety of questions can be put to the computer, with answers forthcoming promptly that would otherwise have bogged down the engineers in a morass of laborious calculations.

The easiness of the "fix" was as welcome to Sam Phillips as was the remarkable speed and ingenuity of the investigation. All that was needed was the elimination of the bellows, leaving a simple uninterrupted tube, looped for flexibility.

The first manned spaceflight of a workhorse Apollo-Saturn vehicle could now proceed on schedule.

iii

Phillips aimed his sights on the month of October 1968 for the historic first test of men aboard the Command Module being readied at Downey. Plans for Apollo 7, as it was designated, did not call for the inclusion of the LM, the mission being Earth Orbital, whence a speedy return to splashdown could be accomplished in the event of serious trouble with the Command Module.

The LM, as stated, was not ready anyway, but it should be reiterated that the biggest problem was not of the manufacturer's own making. It was the added weight inherent in redesigning the craft against fire hazard, as directed by NASA after the Grissom tragedy.

In other respects, the LM was considered definitely spaceworthy, having achieved 32 out of 33 mission objectives during an unmanned flight in January, on

Apollo 5. The single missed objective resulted, you might say, from "overdesign." The Bug had separated itself properly in Space from its Saturn IB booster, and the crucially important Ascent engine, built jointly by North American Rockwell and Bell Aircraft, had performed perfectly. But a discrepancy showed up in the equally critical Descent engine, built by TRW. It ignited on command but shut down after only four seconds, not because there was anything wrong with it, but because its electronic brain was "overcautious." In effect, its brain behaved like a nervous horse that shies at a piece of paper blowing across the road. The common sense of Gene Kranz, of whom the world was to hear more in serious crises ahead, and who was on duty at the time as flight director, deduced that the brain was overreacting to some minor variation of normal engine performance. Kranz restarted the engine by radio command. The mission continued with no further trouble, nor was there any difficulty later in incorporating a "fix" for future missions. Parenthetically, the Saturn IB booster used on this Apollo 5 mission, and which behaved well for the LM test, had been originally scheduled to carry Grissom, Chaffee, and White aloft in 1967.

Schirra, Eisele, and Cunningham would be riding atop a similar workhorse Saturn IB before the next crew would attempt the first maiden flight atop the full, giant Moon-Stack, Saturn V. These three had been practically living at Downey with the Command Module assigned to them, watching hawkeyed over the shoulder of John Healey, the bird's aforesaid den mother, while it emerged from Bill Bergen's extensive spacecraft modification program at Space Division.

iv

Healey told me something of what it was like when he first met the critical gaze of the crew whose lives rested in his hands.

"I knew what was going through their minds," he said.

"They weren't buying Healey on anybody else's say-so. The stakes being what they were, I was guilty until proved innocent, and the answers to their questions had better be good. They particularly wanted to find out whether or not I would be amenable to their suggestions. Here was where I think I avoided the pitfall of sounding like too nice a guy and losing their confidence right off. I said in effect: 'I'll welcome any suggestions you make, and I'll act on them immediately. With this proviso. If the facts finally convince *me* that you've gone for a bad idea, I won't hesitate to overrule all three of you, regardless of your opinion and my respect for you. You've got your job. The hardware is *mine.*'"

"How did it work out?" I asked.

"I think I saw Wally Schirra relax perceptibly," he answered. "Anyway, we got along fine."

Sharing the avid curiosity and concern of Schirra's crew and astronaut Frank Borman, NASA's "buy-off" representative at the Downey plant before Dale Myers sent "Wally's Ship" (as it was called) on to the Cape in April 1968, were Dr. George Low, Apollo program manager, Houston, and his staff of specialists. These gentlemen had been exerting their best efforts toward making a replica of the redesigned spacecraft catch on fire, or flunk the requirement that the new hatch open in seven seconds instead of its predecessor's ninety seconds, or succumb to some other fiendish stratagem that would flush out into the open any hidden, lingering defects. According to all indications, however, North American Rockwell's artisans had finally forged, out of blood, oaths, sweat, and bulldog tenacity, a flight article that would do the job. Faultlessly.

For nearly a year after picking up the baton from Joe Shea, Low had been a stranger to a full night's sleep or a meal at home, as he kept abreast of the progress of Apollo contractors and subcontractors, firsthand, by constant travel. Once, when asked to identify himself and his place of residence, he replied:

"My name is George Low, and I live in an airplane."

Neither was there any respite for him, back in Houston, in the daily search for more answers to the question that consumed everyone: "Has there been any stone, or pebble, left unturned to insure a safe mission?"

The launch complex people at Cape Kennedy—dormant for nearly two years since the last manned flight, Gemini 12—experienced a quickening of the pulse as the time approached for a "live one." (The superstitious can speculate what might have happened to a Gemini "13.")

Mixed with anticipation were emotions of acute anxiety—a busted mission, or worse, fatalities, would almost certainly shatter completely the waning support for Apollo in a troubled presidential election year. Apollo's prime architect, perhaps, and certainly its most powerful supporter over the years, Lyndon Johnson had withdrawn from the political race, harassed by purple thunderheads at home and abroad, and by heavy drains on the Treasury from the floundering war in Southeast Asia. His successor, be it Nixon or Humphrey, would sorely need a big win to keep the show on the road to the Moon, under a new administrator (Webb was to resign in October, leaving Dr. Thomas O. Paine as acting administrator of NASA until his permanent appointment by Nixon in March 1969).

V

Apollo 7 rose flawlessly from ill-fated Pad 34 on October 11, 1968, and completed 163 revolutions of the Earth during its eleven-day flight, hailed by a vastly relieved George Low as "101 percent successful," several extra, nonscheduled mission objectives having been achieved. And what problems there were having been minor, George Low and North American Rockwell had every reason to feel vindicated. For Lee Atwood, in particular, the sun once again shone bright.

We know now, however, what the consequences might have been had the same emergency arisen that was to befall Apollo 13—a complete failure of the primary electrical power and life-support systems in the Command and Service Modules.

"Could Apollo 7, in the absence of the LM for backup, have reentered from Earth orbit after an identical failure," I asked an expert, "without risking a firing of the main engine in the badly damaged Service Module, or, failing that, its small thruster engines?"

"Not a chance," he said flatly. "The Command Module has no way of slowing down enough to descend from Earth orbit in time, unless you use the Service Module's engines. The Command Module's thrusters are for attitude control only, not retrofire. So we'd have been forced to try burning the SPS engine in the damaged Service Module, or if it failed to burn, the reaction thrusters, regardless of the risk. Simply no choice. It's anybody's guess what might have happened *then.* To date we just don't know, and we may never know."

Certainly in the actual event on Apollo 13, no one advocated such a gamble—except as a dire last resort.

Shortly after the celebrating had subsided, Chris Kraft and his flight planners at Houston began putting the arm on Sam Phillips to shoot the works on the upcoming Apollo 8 mission, originally scheduled for December as another Earth Orbital job, in which the crew would check out the LM. Since the LM was not yet available, they argued, why not go for the lunar orbiting mission hinted at as a possibility by Phillips in a press conference back in August? Phillips had spoken in the context, of course, of a successful Apollo 7 mission that had now become an actuality. Cautious but confident optimism prevailed. On November 12, 1968, acting Administrator Paine voiced the considered judgment of Dr. Mueller and Phillips in a dramatic announcement:

"We are now ready to fly the most advanced mission for our Apollo launch in December, the orbit around the Moon."

One might ask now, in comfortable retrospect, why Paine, in arriving at his conclusion that men could be sent around the Moon at acceptable risk without the LM "lifeboat," did not attach more weight to the possibility of a complete main power failure in the fuel cell system?

The answer is that such a drastic prospect, to his advisers, was so unrealistic that crews were not even required to simulate countermeasures in their training. Any more than they simulated emergency measures to cope with a pulverizing collision between the spacecraft and some huge meteorite, or a similar "act of God." Near-total incapacitation of the Command Module fell into that extreme category, hence a waste of the crew's precious time in a tight training schedule. Had any engineer forseen even a remote chance of a catastrophic failure of electrical power, despite all of the redundancy of backup systems, he would have recommended, "Don't go at all." Phillips would have been the first to concur. Until the LM was ready.

Accordingly, since you would not go at all without great confidence in the demonstrated performance of the Command and Service Modules, you were justified in deciding, as Paine did, to go without the LM; for if you overplanned against the worst possible exigencies, it would be consistent to postulate that the LM's systems might fail also, and then where would you be? You'd stay poised on the tip of the diving board indefinitely.

In the light of all facts then available, Paine's decision to permit men to spend Christmas up the creek behind the Moon without the LM "paddle" does not seem rash. Bold, but no rasher than the Mercury and Gemini flights that had to be risked against inevitable and more numerous unknowns.

True, the decision raises goose pimples in retrospect. But in 1968 there were a host of other goose pimples for company. In fact, they were the name of the game, and will continue to be, until that day, perhaps sooner than we think, when Space flight becomes "routine."

vi

Comparatively recent though it is, the blastoff of Apollo 8 on December 20, 1968, was preceded by a buildup of suspense that is difficult to relive in one's memory today. The humble and the great from far parts of the globe converged on Cape Kennedy, all eyes turned toward the mighty Stack gleaming on Pad 39A, its breath steaming. Every TV screen in the land and overseas began glowing overtime.

Dr. Kurt Debus, Center director and father of the vast launch complex, confessed that he would have found the tension unbearable, had he not already been through so many nerve-racking countdowns—none quite like this one, however, where astronauts who had previously confined themselves to relatively safe excursions in Earth orbit, no further from splashdown than New York is from Washington, D.C., would venture a quarter of a million miles out into the void. And for the first time they would enter the gravitational field of another body. Escape from that alien lunar gravity, hence crew survival, would hinge upon the reliability of a single engine, built for North American Rockwell by Aerojet-General Corporation (a division of General Tire & Rubber) at its Sacramento and Azusa, California, plants. More than any other, the single question that commenced to dominate the suspense, centering on that Aerojet engine, became:

"Will the damn thing start?"

It had started thousands of times in ground-test firings, and eight out of eight times on Apollo 7, when it had been called on for course corrections and the final de-orbit, or "go home" burn from Earth orbit. But this time the de-orbit would occur behind the Moon.

It is safe to say that no one on Christmas Eve was watching the Apollo telecast with more moisture on the brow than four intimately involved Americans and one Yugoslavian. They were Herman L. Coplen and Dan David who, successively, had headed up Aerojet's rocket engine program for the Service Module of Apollo; the man they had to satisfy at North American Rockwell, Bob Fields, responsible for the SPS (Service Propulsion System) of the Service Module; "Mike" Vucelic, a key design engineer from the inception of the Apollo spacecraft and an immigrant from Yugoslavia, and his father.

The elder Vucelic, distinguished in his own right as an engineer and railroad builder who had driven tunnels through Yugoslavia's mountains, had paid a visit to his son at Downey while final touches were being put upon Apollo 8. He was impressed but skeptical after his son had given him a thorough briefing on how the amazing vehicle would cicumnavigate the Moon.

"Iss good locomotive," he said, "but where are the tracks?"

Months later, after an unbearably long radio silence behind the Moon until Andy Vucelic and the world heard Borman's welcome "There *is* a Santa Claus," and Apollo 8 started for home, Mike rushed out and sent a cablegram to his father:

"THE TRACKS ARE THERE."

THE ACID TEST (1969-1970)

"Many of us still can't believe that the goal we set out to reach in 1961 has been achieved."

George M. Low, Manager
Apollo Spacecraft Program Office

When I embarked in April 1969 on extended visits to beehives of Apollo activity in their final fever of preparation for the lunar landing, at Houston, Huntsville, Cape Kennedy, and other centers, I of course did not have the background of the preceding chapters. It was all fresh, unplowed soil. In a personal sense, the trip was "Chapter I," but to have written it as such would have entailed constant backtracking. Now, with a better-informed perspective, I trust, which I can share with the reader, I have caught up at last with my story, in the climactic stage of the great human event scheduled two months away, July 16: Apollo 11.

Many of my preconceptions were soon to be modified, as I checked into a cool motel room at the King's Inn, mercifully air-conditioned against the tropically fierce, muggy heat in which the adjacent Manned Spacecraft Center was sweltering alongside Clear Lake, where the astronauts water ski (it is neither clear nor a lake and is filling up with silt), 25 miles southeast of Houston. Among the preconceptions:

"A NASA public relations functionary will sit in and monitor all of your interviews."

John McLeaish, who provided me with escorts for a score of interviews, did not require any of the escorts to be present. It was entirely up to me. Those who deal with the public and the news media, like meteorologists, traditional butts of humor and scorn, can seldom "win." About the best they can hope for is to avoid ruinous blunders. NASA's performance, in keeping the public as fully informed about Apollo and its builders as possible, I was to find better than most, stemming from policies of openness that originated with Walter T. Bonney, under Keith Glennan.

"How well satisfied are NASA executives with North American Rockwell's overall performance?"

Everyone from Bob Gilruth and George Low on down told me that North American Rockwell had wound up doing an excellent job, as fine as could be expected of any company in America.* One high official thought that the contractor had had to overextend its manpower in grappling with so many major Apollo commitments. Another believed that McDonnel Aircraft (since merged into McDonnell Douglas), which built the spacecraft for Mercury and Gemini, should for that reason alone have been given a similar responsibility for Apollo. It obviously had the bulge, he said, on experience.

"NASA has taken the bows, but private industry really designed and built the Stack."

Actually, it had been an unprecedented blend of mutual creativity, as was abundantly clear here at Houston. Whereas the government went nearly all the way in delegating responsibility to a prime contractor for the entire systems for the Atlas ICBM, the Thor, the Minuteman, the B-70, the C-5A jumbo transport, and the F-15 advanced interceptor, NASA's own, in-house designers

*Gilruth made it official in a letter to Space Division, part of which reads as follows: "The successful fulfillment of the national goal of landing men on the moon and returning them safely to earth in this decade was in large measure due to the outstanding management and technical skill reflected in the design, development, and manufacture of the Apollo Command and Service Modules."

pretty much *told* industry *what* they wanted built for Apollo and Saturn. This represents a partial reversion to the old Army-Navy "arsenal" concept. It worked because NASA was blessed with a concentration of creative minds nurtured at Langley Field under the aegis of its predecessor, the NACA, which I was now encountering in the flesh—and at Huntsville, under Von Braun.

"The astronauts are the stars of the show."

Deke Slayton, Jim McDivitt (who has succeeded George Low in the upper echelon of MSC), and other astronauts were the first to tell me that the flight directors in Mission Control under Chris Kraft and Gene Kranz (and of course including them), had to "live" through every mission almost as realistically and as professionally as the Apollo crews themselves—right up the limits of man's physical, mental, and nervous systems. In the eyes of the initiated, those flight controllers were fully entitled to share top billing.

"Civil servants such as Gilruth, Kleinknecht, and Faget, originally schooled in pleading for research funds rather than in *spending* money for huge procurement programs, must be unpromising material for top management slots."

All of the key NASA managers I met, having survived the shakedown years, impressed this observer as rather extraordinary men—modest in spite of brilliance, drivers, and perhaps most characteristic of all, conscientious to the point of fanaticism. They were able to make *decisions.*

ii

Modesty was certainly the appropriate word to fit Kenny Kleinknecht, number two manager for the Command and Service Modules, a bear-sized, large-nosed, mild-mannered, and impressive man. Comfortable as an old fireside slipper, on a first meeting, he nonetheless projected a formidable presence. He frustrated my immediate purpose by talking about everyone else but himself, minimizing his role in Apollo's evolution, eagerly thrusting at me other names besides the inevitable Max Faget—unsung, or at least less well-known worthies like Bob Piland and "Cadwell" Johnson.

Kleinknecht seemed oblivious to the fact that his peers rated him a Rock of Gibraltar. He was also oblivious to the clock, having scheduled me at an hour when most men go home. When I had demurred to his secretary, she said:

"Don't give it a thought. He may be here 'til midnight."

Although I failed to steer him into the subject of Kleinknecht, he did impart to me a bit of his own philosophy that, incidentally, meshed well with Faget's. We had been discussing quality control and reliability.

"You can improve a good system into real trouble," he said. "If it's doing the job, *leave well enough alone.*"

A recent example immediately crossed my mind—the shutdown of the LM's descent engine (described earlier) because of oversensitive electronic sensors. And later I recalled the mission on which Apollo 13 suffered the explosion related to a short circuit in the wiring of fans in an oxygen tank, installed there to keep the oxygen stirred up. It now develops that the fans may not have been essential in the first place.*

But according to Kleinknecht's reputation, for every "overcomplexity" which got by him, there must have been a hundred that he weeded out.

iii

In George Low's spacious office, with its uncluttered desk, the first thing that caught my eye was a very much cluttered side table. It resembled a handyman's workbench, filled with odds and ends of switches, valves, and other assorted small hardware.

*NASA's final review has zeroed in on a non-spec thermostat switch to the tank's heaters as the presumed culprit, during ground tests.

My host, who could be typecast for the movies as an intellectual-type football coach, about as easy to satisfy as was Vince Lombardi, and who has a sort of devastating calm about him, explained the presence of the gadgets.

"Those are parts that have failed," he said, "or that might be improved. Every week General Electric, as integrator of all the Apollo contractors, gives me a briefing on trouble spots in the hardware."*

He picked up two or three of the parts and told me in minute detail where each had been found to be defective. I scratched my ear in puzzlement. He noticed it.

"Something I didn't make clear?" he asked.

"Yes," I said. "I'm not clear, at your level, why you have to concern yourself with this much technical detail when there's an army of experts at your disposal. Forgive me if I'm asking a dumb question."

"No question is stupid," he said agreeably, "if you don't know the answer. Okay. I'll give you an example."

He picked up another gadget.

"A modification has been recommended in this arm-joint for Neil Armstrong's spacesuit. That immediately raises the question of the time element, this close to the launch date. Can we provide time in Armstrong's tight training schedule for him to check out the change in a test chamber? If not, do I want him to risk finding out for the first time on the Moon's surface that the "improved" arm-joint is giving him trouble, because he isn't used to it?"

His secretary brought us the inevitable black coffee as we returned to his desk.

"That's the kind of decision, small as it may seem, that I can't delegate," he continued, "because only I have the whole picture, especially timewise. It follows that I can't pass on a technical decision unless I understand the problem as clearly as the engineer responsible for the

*Boeing shares the integration function with GE, which concentrates on processing data and preventing "communication gaps."

part. And I'd better understand this one by two o'clock this afternoon when a meeting is set up to reach a decision. Does that answer you?"

"Question withdrawn," I said, smiling, feeling like clicking my heels.

I then asked him about responsibility for the fire on the pad that occurred under his predecessor, Joe Shea.

"It not only could have happened to *any* contractor," he said, "but it could have happened at any time during the many previous Mercury and Gemini flights, in the same 100 percent oxygen cabin environments. It was a case where *every*body shares the responsibility."

He expressed his antipathy to "overinspection" of hardware.

"The worker loses incentive," he said, "if too much of his responsibility is usurped by inspectors. Let the foreman inspect his work first, so he is motivated to do the job right in the first place, for his immediate boss."

I touched on the widespread belief that Houston was chosen as the site for the Manned Spacecraft Center, and Cape Canaveral for launch operations, on political grounds, under President Kennedy and Vice-President Johnson.

"I won't comment on the political factors," he said, "but I will predict that detached historians will conclude that both decisions have worked out well for the Apollo program, geographically and otherwise—and for future programs."

In discussing Apollo 10, now less than three weeks off, in which the LM would separate from the mother ship and descend for the first time close to the Moon, to scout out the July landing of Apollo 11, Low expressed a quiet confidence that his machine was ready. When I left him, once again I experienced that tingling sensation that comes with stumbling onto the clues of a mystery—the mystery that tantalizes the TV viewer when he pinches himself in disbelief at the sight of men actually setting course for the Moon.

George Low's office was one of those sorcerer's chambers where mysteries are unraveled. I had just met another of the sorcerers.

His not-so-mysterious gift? Like Gilruth, he knew what he was doing. He also knew that Armstrong and his crew were staking their survival on that fact. And with peace of mind.

iv

Already well-publicized astronauts not being a first concern of a book about Apollo-Saturn's builders, my meetings with Donald K. "Deke" Slayton, Gerald P. Carr, and James A. McDivitt were brief. And they were relevant to the extent that the race driver's role is intimately involved with the car's builder.

Indeed, a large part of astronaut Carr's job, as he explained it, was to haunt the plants of the contractors. It had been learned on Mercury and Gemini that the prime crew and the backup crew were being forced to divert too much of their time and energy to "midwifing" their embryonic spacecraft during production, at the expense of more urgent operational training and other duties. Therefore, a third crew was now being selected for each Apollo mission, called a Support Crew; Carr was a member of the three-man support crew for Apollo 12.

"We've been able to take a big load off the prime crew," he said, "not only sparing them a great deal of travel time, but study time. From our legwork on plant visits, we can boil down a mass of technical information into salient points and give the first team a condensed briefing."

"Another plus from our visits," he continued, "is in keeping the workers, down to the last welder, reminded of the human stakes under our 'Manned Spaceflight Awareness' program. Meeting live astronauts keeps the girl at a bench from getting impersonal about what may seem to have become a dull, obscure job, when the truth is that the two-bit part she's wiring, if it failed, could cost lives."

On the operational side of things, Carr gave me an example of the cohesion between the astronaut in Space and the flight controller on the ground. When Borman's crew disappeared behind the Moon on Apollo 8, out of touch for over half an hour, ground experts were sweating out the spacecraft's dwindling water supply. They pooled their brains to analyze the problem and be ready with answers by the time Borman reappeared after "cutting the hill," as Carr phrased it.

"The controllers had reached step number 12 on the checklist when radio contact was restored," Carr said. "Bill Anders called down that he'd been working on the checklist and was at step *12*—exactly the same point. How's that for saving lost motion?"

V

Deke Slayton, who carries the heavy responsibility for directing crew training and flight operations, has to be the most frustrated astronaut of them all. Deke might well have become the dean of the spacemen, if it had not been for the slight heart murmur discovered before his first scheduled flight, which resulted in a borderline decision by the medicos to ground him permanently, early in the program.

That he was obviously still itching to fly at the drop of a paper clip spoke well for his faith in the builders and the prospects for success in the imminent "big ones." Slayton is the archetype of analytical skeptic who cannot be lured aboard a machine unless he's been given a good one. Thorough as they come.

"Have you given up hope of pulling a mission?" I asked.

"I don't think it would be fair, this late in the game," he said, "for me to bump a younger sport off a crew, but if an opening comes up for a combination scientist and astronaut crew member, I think I've done enough homework to qualify on both counts. No, I haven't given up."

Meanwhile, like Bob Kiputh, Yale's famous swimming coach during a long reign of national supremacy in which he never went near the water (it was rumored that he couldn't swim), Deke Slayton, wise and cool-headed, has done a whale of a job without going near outer space.

vi

The evening I spent in the home of astronaut James A. McDivitt came about accidentally as the result of the latter's perfectionism in constructing a marguerita cocktail.

I had just arrived at the home of one of McDivitt's neighbors, Jack Waite, and his tall, very beautiful wife Aletha (he is North American Rockwell's representative at the Manned Spacecraft Center). The highlight of the evening was to be a telecast in color of the first rendezvous and docking of the Apollo spacecraft and the LM, on Jim McDivitt's recent Apollo 9 mission in earth orbit.

Waite answered a knock on the door and introduced to me a sturdy lad, one of the McDivitt's four children, who stated his business tersely:

"My old man's making margueritas, and he's fresh out of Triple Sec."

In a flash, Waite fetched a bottle of Cointreau from his bar, which the boy reached for.

"Not so fast, son," Waite told him. "We'll deliver this personally."

As the four of us walked the short distance to the McDivitt's, Waite informed me that margueritas á la McDivitt were "nominal" (astronaut lingo for perfect) and that in any event the McDivitts were a family I'd enjoy meeting. So they were.

Mrs. McDivitt, resembling an elfin, 15-year-old version of Shirley MacLaine, greeted us in the kitchen, where plates were set out in the breakfast nook for her hungry brood, two of them in high chairs. Smoke wafted through the window from steaks barbecueing in the backyard. Shortly our host appeared, a small-boned, pale-faced man clad in a University of Michigan sweatshirt and bearing sizzling steaks, which he delivered to the older children before shaking hands.

Guarded at first with a stranger, as I had been forewarned by Waite, he gradually warmed up along with the margueritas he generously dispensed, and with talk of flying. I asked him how he got into aviation.

"Reading," he said. "As a kid, I read all the pulp magazines, everything about flying I could lay my hands on, including that book of yours. I got so presold that I entered Air Force flying training without ever having been in an airplane, not even for a ride."

This struck me as highly unusual, like buying a car without driving it out of the showroom, and it said a lot about McDivitt. Having made up his mind in advance that he wanted to be a career military pilot, apparently the question of whether or not the airplane would cooperate with his decision became incidental.

Outwardly, he was relaxed, deliberate, but I got the impression of a tightly coiled steel spring inside him, a controlled tension, that would drive him unmercifully toward a chosen goal.

There had been speculation in the press that the crew for the approaching Apollo 10 mission, Stafford, Young, and Cernan, was being pushed too hard in its arduous schedule of preparation. I asked McDivitt if he thought Apollo missions were taxing human capacities too far.

"I'd say we're going fairly close to the limit," he said, "both in preparation and during the mission if problems pop up, especially for the Command Pilot who has to make the decisions. No matter how much help you get from the ground, and the controllers have been magnificent, it's still advisory. Technical advice from contractor reps helps a lot, too. But you can't have overlapping decision making by committee. When you're up there, with rare exceptions, the onus of decision making is squarely on you."

"Were you close to human limits on your Apollo 9 mission?"

"Apollo 9 was *extremely* demanding."

He didn't elaborate, but I'd heard from other sources that matters had not been made any easier by the messy nature of the illness of a crew member during part of the mission.

"When you're aroused from a sound sleep up there," I resumed, "how long does it take you to reorient yourself, come wide awake?"

"I'd say a minimum of 60 seconds before I'm really hacking it again—aware of where I am, what day and what time, and what phase of the flight plan we're in."

"In an emergency, could that turn out to be 55 seconds too long?"

"Possibly," he said, "if all three of us were sound asleep at the same time. Which we didn't use to allow. Then we concluded that there are times when it's unwise to try to keep one man on duty, if it's a phase of the mission where all *three* astronauts are badly in need of rest. Mission Control down there never sleeps, of course, and they're monitoring every system."

He glanced at his watch.

"About time for the TV special," he said, and turned on his set.

To my amazement, the picture was in black and white.

"You're not going to watch it in color?" I asked.

"Can't afford a color TV set," he answered. "Too expensive, with the size of this family."

He was quite serious, as Waite told me afterward. McDivitt was one of the majority of astronauts who scrupulously avoid capitalizing financially on their prominence, beyond pooled royalties for publishing rights, for example.

Thus on an evening in which America watched the most recent Space spectacular in color, the star of the Apollo 9 mission, and perhaps the most interested single spectator, watched the historic broadcast in black and white because he preferred to spend every possible rare moment with his family to watching on a neighbor's set.

After voicing our good-nights, the Waites and I passed the open garage, crammed with family gear—bikes, an outboard boat, water skis, and gardening tools. Waite pointed out some flourishing young live oak trees bordering the lawn.

"Jim planted those," he said. "Done a good job of nursing them along, too."

"It figures," I thought to myself. "Roots. Deep roots. That's McDivitt, home from Space where there aren't any."

vii

Eugene F. Kranz, flight director and number two to Christopher Kraft, head of the Flight Operations Directorate, joined me for lunch at the King's Inn, an event that was unusual in itself. Normally, Kranz doesn't have time to pause for lunch, even at his desk.

Aged only 35, chunky, of medium height with blond, close-cropped hair, Gene came as close as anyone I've met to being a human dynamo. He obviously got a charge out of the torrid pace of his job, thriving on it with the addiction of a narcotics user—the kind of man who seems to want to cram a minute of living into a second. Whether this was the cumulative result of his job over the years, or a prerequisite for the job of flight director, it is hard to say.

One of the first things he jolted me with was the casual observation that for the Apollo 10 mission he'd be carrying approximately 2,500 different "canned" emergency procedures *in his head.* Memorized. That way, he explained, he needn't risk wasting even a few seconds looking up the right solution in the book, where a couple of seconds might make a big difference.

"There's a time lag of a second and a half to begin with," he said, "before a controller's voice reaches the spacecraft near the Moon. So you and the astronaut might be on the mike at the same time, drown each other out and have to start all over. On a Mars mission, you can be bucking a lag of 25 *minutes.*

"We select controllers as young as we can," he went

on. "Right now we've got a good man only 21 years old. We're forced to be extremely choosy, because the candidate will be up for membership in maybe the most exclusive club in the world. Ideally, a flight controller has a few bucks' worth of fighter pilot in him and a lot of the test pilot and the control tower operator at O'Hare in his makeup, all in one package of dry ice.

"There's a premium on stamina. We work three eight-hour shifts on an Apollo mission, but it really comes out twelve hours—two getting warmed up and clued in to come on duty, and two afterward to turn loose of the reins for the fresh man and to unwind.

"When a controller makes a mistake, I'm not concerned so much with *what* he did as finding out *why* he did it. Was it overfatigue, for instance? One of our guys drove his car into a brick wall just before a mission and had to be replaced—not because of injuries, but because there was no explanation for what came over him. Too much strain, probably.

"What particularly concerns Chris and me are errors that fall into a pattern of repeating themselves. Like a controller failing to check fuel reserves. Why? Why do intelligent people run out of gas on the freeway? Often that kind of mistake comes during the unexpected, after there's been an interruption in the checklist. So we postulate that departures from a sequence of procedures are a booby trap, like the man interrupted during dictation of a letter who had better ask 'Where was I?' or he'll leave out something important.

"Most of all, Chris and I look for coolness. We can find out plenty ahead of time by exercises in which we simulate a phase of the mission and crank in all kinds of awkward emergencies. The controller comes out wringing wet. And we want the controller to find out just what it's like for the astronaut. We put him in the same type simulator as the astronaut's spacecraft and make him experience very closely the same physical discomforts of the cramped environment, so that in the real thing he won't issue instructions to do something that's clumsy or impossible for the man on the couch in Space."

Gene was chewing in a desultory manner on a cigar.

"Are you off cigarettes?" I asked. He nodded glumly.

"I had to do *some*thing," he said. "On long missions like eleven days during the last one, Apollo 9, I was just about incapacitating myself on cigarettes. I ought to quit cold, but I'll see how cigars work out on Apollo 10."

After lunch we went over to Mission Control, where a small army of shirt-sleeved controllers and plant reps were glued to their consoles in one of the endless simulations Kranz had been telling me about, for Apollo 10. He explained how the flight director conducted the members of the orchestra and how, in addition, he could in a matter of seconds, by phone, call upon engineers standing by at contractors' plants all over the country for analysis of a malfunction.

"It's an odd thing," he observed, "but the man who designed the piece of hardware, or the system, is seldom a decision maker. He can tell you exactly what went wrong, but not necessarily what you ought to *do* about it. That takes an *operator.* And that's what we're paid for—and what we look for in a controller."

It wasn't until later, just before the Apollo 11 lunar landing, that I was to meet Kranz's boss, Chris Kraft. At the moment I found myself wondering, is it really necessary to meet Kraft at all? If Kranz is his handiwork, then why not settle for the adage, "By their works shall ye know them"?

After he had further briefed me on other sections of his symphony orchestra—the tremendous backup of IBM computers and the global network of communications and tracking stations at the conductor's disposal—Gene and I returned to the blazing heat outside the Mission Control building. I stared about us, disoriented.

"This isn't the way we came in, is it?" I asked. "Going to your office?"

Now it was his turn to look confused.

"No, we walked over here from my office," he said. "Remember?"

I didn't. So engrossed had I been by his conversation, and by the effort of trying to keep up with a brain that seemed to be revolving at many times the r.p.m. of my own, that I had walked with Kranz 200 yards from one building to another without being remotely aware of it.

That is when I first started ruminating about Borman's reference to riding toward the Moon on the shoulders of giants—no two of them quite alike. Fascinating men.

viii

Two weeks remained before the May 18 blastoff of Apollo 10, but so great was public interest in this final rehearsal before the actual landing, that sight-seeing early birds were already homing in on Cocoa Beach, Florida, thronging the motels and beaches near Launch Complex 39. They knew that the mission would be the riskiest yet attempted by far, with Tom Stafford and Ed Cernan scheduled to leave John Young alone in the Command Module "Charlie Brown," while they flew "Snoopy" down to within 9 miles of the Moon's craters.

I too checked in several days early at the Cape, after spending a week at Huntsville to interview Von Braun and his fellow rocket-builders, as related previously. It might be appended that Von Braun did not share the concern current in many quarters that Moscow might still try to beat us to the punch before Apollo 11 in July. He gave the Russians credit for enough booster power to assay a direct ascent to the Moon but believed them to be backward in onboard guidance and navigation systems, in which technology the United States had surpassed them, with the Soviets being constrained into greater reliance on ground stations.

"However," he had conceded, "if we should run into any serious difficulties on Apollo 10 that slip our schedules into next year, it might be a different story."

The contagion of excitement at the launch complex

infected and seldom left me, as I made my rounds, with the final countdown to May 18 already in progress. One who clearly enjoyed the tension was, in a sense, everyone's host, Dr. Kurt H. Debus, Director of the Kennedy Space Center.

"The strain on the managers here," he said, "is such that you've got to *like* what you're doing. If a man doesn't enjoy the pressures of final checkouts and countdowns, I advise him not to fight a losing battle, but to try something else with apologies to no one."

A small, quiet, courteous man, Debus is the builder charged with assembling Apollo-Saturn's major segments after they arrive by air or by barge, checking out all systems and launching the Stack safely away from the immediate vicinity, almost as gargantuan a task as building the Stack in the first place, or of coordinating the veritable mountains of paperwork that accumulate during its gestation. As Debus put it:

"When the weight of the paperwork equals the weight of the Stack—7.5 million pounds—it's time to launch."

Accompanied by a sharply well-informed girl guide named Angela, assigned to NASA by Space Technology Laboratories, I had my first look at the largest "house that Debus built," and the second largest in the world, the famous VAB (Vehicle Assembly Building), capacious enough to accommodate simultaneously four Stacks, so as to stay abreast of launch dates as close as two months apart. A cool breeze blew through the open doors of the mammoth structure, a welcome relief from our ride in the stifling staff car, equipped with what Angela called "100 percent NASA air-conditioning" (open windows) and "100 percent NASA power steering" (brute strength).

Expecting to ride up to the 34th floor, where I could look down on Apollo 11, now nearly ready to be rolled out the door on its mobile launcher for the crawling first step to Pad 39 en route to the Moon, I was surprised when Angela took me only as far as the fourth floor. After she had shown me the lower stages of the partially assembled Apollo 12, she approached the Down elevator.

"Aren't we going up to 34?" I asked. An uncomfortable expression came over her face.

"I—I'd have to get you an escort," she said.

"Can't you take me?"

She turned pale.

"I've never been able to get up the nerve," she said. "I'm afraid I'd pass out on you."

After a fruitless effort to find another escort for me, and aware of my obvious disappointment, she struggled painfully with herself. Suddenly she squared her shoulders, ushered me into the Up elevator, and pressed the button for level 34. Her color approached a pale green as we zoomed upward, arriving with a rush opposite a grated platform through which we could look straight down into a yawning chasm. The effect even on me, an aviator, was breathless, and it was almost as dizzying to gaze upward at a 250-ton capacity crane at the 525-foot summit of the cavernous interior of the VAB. Just below us loomed the tip of the gleaming white shaft destined for the Moon landing.

Angela had remained in the elevator, temporarily frozen. Then her expression changed and she stepped out unhesitatingly and joined me, right at the outer railing of the giddy platform.

"It's all right, now," she said, gripping my hand nonetheless.

Close proximity to that triumphant pinnacle of engineering, the A-11 Stack of so historic a destiny, gave me a chilling thrill. But it vied with another thrill—the simple triumph I had just witnessed of a frightened girl's spirit over a paper tiger in her mind. Of such is the kingdom, also, of astronauts, and of all who meet their own confrontation with the Unknown.

ix

General Sam Phillips was finishing up a TV recording session in a room just off the Launch Control Room, fully manned by 400-odd experts, when I arrived for our appointment. He looked awful. Really ill. Yet he was answering questions concisely, a very cool customer, as he deplored the tendency to charge off a $24 billion price tag entirely to "The Moon," when so many of those billions, he said, represented priceless investments that would pay off in other programs both in Space and on Earth.

"Hasn't happened to me in ten years," he apologized to me as soon as he was free, "but I've picked up a virus. A dandy."*

I offered to cancel, but Sam insisted on accommodating me if I didn't mind coupling the interview with a visit to the biomedical offices, where he was due for shots. In his car, which he drove with shivering hands, and in snatches at the dispensary, where Mission Director George Hage was also in for a checkup, he filled me in on latest developments in the countdown and on Stafford's crew, which was in good shape, but to me it is the human sidelights of that meeting which merit recollection here.

Instead of going to bed, which Dr. Berry had just recommended, Phillips had decided to stay on the job, following the countdown personally all through the night and the next and final night, continuously on call even when catching forty winks. Along with this, he planned to (and did) adhere to a taxing schedule of conferences, press interviews, and other commitments. Every major decision now rested on him, and it was on him that his boss, Dr. George E. Mueller, would largely rely. Fever or no fever, a couple of hours' sleep here and there would have to suffice until liftoff.

I thought about this kind of inner toughness when I awoke the next morning shaking with an acute attack of the 24-hour flu, apparently passed on to me by Phillips. Woefully weak, and grateful that no decisions more

*I didn't learn until months later that he had pneumonia.

immediate than ordering breakfast faced me, I toyed with the idea of canceling my eleven o'clock appointment with Dr. Mueller. By 10:30, however, the example of Phillips' willingness to use up the last drop of himself prevailed.

x

Dr. Mueller was registered in Room 110, but I thought I'd come to the wrong place when I crossed the courtyard from my bedroom at the Holiday Inn and found, not the suite of a VIP, but cramped quarters, and a lean bespectacled figure in an undershirt, sitting on the unmade bed and talking to a lady. Shortly she folded her notebook, turned off a tape recorder on the bed, and relinquished Dr. Mueller to me. He apologized for his dishabille.

"Not enough hours in the day," he said, then reached for a buzzing phone.

"Another Hold? . . . Wetting the wick on the ECS?" he said, frowning. "Oh boy." And again, after an interval, "Oh boy . . . well, keep at it, and let me know."

"Anything serious?" I asked.

"Nothing too serious," he said.

Quick of eye and quick of wit and humor, Dr. Mueller was completely in keeping with his unpretentious setting, informal, a no-nonsense, no-lost-motion type. His credentials, I knew, were formidable, as civilian architect while at STL (Space Technology Laboratories) of our ballistic missile program in the Fifties, and as a NASA planner who had briefed President Kennedy way back in 1963 on the details of a projected Apollo 10 mission now about to be flown without a single major alteration from the original concept (as Phillips pointed out to me).

Again, let me leave you with a human sidelight rather than recounting the mission shoptalk that was of so much concern to both of us at the moment, with liftoff scheduled for the next day, Sunday afternoon, at 12:49. A correspondent poked his head through the door and inquired whether Dr. Mueller could squeeze him in somewhere for a few minutes. The latter scrutinized his list of appointments, scowling and shaking his head, then brightened.

"Why, yes," he said. "I've got a free spot for a whole hour between noon and one o'clock tomorrow . . . my only free time for two days."

The correspondent looked dumbfounded.

"You aren't going to watch the launch, Doctor?" he asked.

True to the Absentminded Professor in his background, NASA's eminent Director of all Manned Spaceflight had indeed momentarily lost the forest on account of the trees. Recovering, he laughed harder than any of us.

"Can I use this?" I asked him.

"Sure," he said. "Why not?"

xi

Many times before, watching television without too much emotional reaction, I had seen astronauts step from their crew quarters and enter the van for that "Last Mile" to the pad. But when I stood only a few feet away from Stafford, Cernan, and Young at 9:42 the next morning, and saw them smiling bravely at the crowd and waving before they climbed aboard, accompanied by technicians carrying their life-support packs, I felt tearful. Somehow, the finality of their commitment, as the van rolled away, struck home in full force. There was a lump in my throat.

Likewise, I found that there was little comparison between watching a blastoff on television and witnessing the real thing from the Press Site, about three miles distant from the pad. The Stack lifted off seven-thousandths of a second late, just short of perfection. From my notes:

"I've seen the miracle—a magnificent, shattering, terrifying sight and sound, on a scale with the fury of the elements, channeled into an execution of man's will. More hundreds of millions of horsepower than were available in the entire world as recently as 1900 had just been unleashed in one pillar of fire. Simply beyond description."

So perfect was the mission as a curtain-raiser that Armstrong and Aldrin's landing on the Moon two months later, on July 20, seemed largely a replay of Apollo 10, almost anticlimactical, with success so palpably probable.

Having caught the live show at the Cape, I decided to witness Apollo 11 from a different vantage point—Mission Control, Houston. Like the builders, I prayed for the near-perfection that every Moon visit demands, and which could terminate my story in an undiluted blaze of glory. But Apollo 13, waiting in the wings, was to intervene before I went to press.

10

A LONG WAY FROM HOME

"Go as far as you can see. When you get there, you will be able to see further."

Thomas Carlyle

It was the day before the liftoff of Apollo 11, and at Houston elaborate working facilities were in readiness for the multilingual worldwide press. At the moment, the bulk of the latter were adding to the crush of dignitaries and tourists at a Cape Kennedy which was reported to be slowly sinking under the weight, but the media would begin migrating en masse after the launch to report the denouement here at MSC.

Having the place nearly to myself, I determined to exploit the lull in an attempt to penetrate the inner sanctum of Mission Control. This, of course, was not going to be easy for anyone lacking demonstrable qualifications as a participant in the final countdown under way.

Gene Kranz being in the Control Room, which he could not leave, I asked his secretary Lois if I might go see him. A sympathetic conniver, she picked up some papers and a telephone extension cord that Gene wanted sent over, and bade me follow her.

"I can't promise how far you'll get with that press badge alone," she said, "but if we're together it may help."

Like Ulysses clinging to the wool of a sheep's belly, I successfully ran the gauntlet of a succession of guards until we arrived at the final barrier to the Control Room. Here the guard explained that only Kranz personally could admit me. Soon Lois reappeared with my crewcut-headed friend. Pale and tense with preoccupation, but polite, he got me a green badge and ushered me into the cerebellum of Apollo 11, whose destiny he would direct. The charged atmosphere around the consoles was so thick you could chin yourself on it.

Initially I regretted my intrusion—Gene was too busy for distractions—but fortuitously his boss Chris Kraft was present and free to talk, and I had long been eager to meet this household word from Mission Control. An originator, as well as a veteran from the Mercury days, of the art of planning and controlling spaceflight operations, Kraft looked and talked the part of a flight director, soberly self-confident, outspoken in his opinions. By contrast, two young controllers he pointed out to me, scholarly bespectacled specimens both, did not at all look the part nor resemble their athletic-type mentors Kranz and Kraft. Yet they had been assigned to hairy responsibilities bearing on a reliable descent to the lunar surface.

Behind the deceptively mild exterior of these two, Kraft explained, resided a rare ingredient which he elaborated upon under the incongruously unscientific term of black magic.

"Those two controllers can't just rely on pure math," he said. "They need nerve and an intangible ingredient, a sort of creative intuition in arriving at good answers. I'm not talking about blind hunches, but there *is* an element of black magic in the kind of human judgment that has to be stirred into this computerized recipe for controlling a mission. Often what the computer tells you to be the ideal approach is correct as far as it goes. But a good controller knows when to settle for less, if for example choosing a less desirable approach is a sure bet, measured against a more desirable approach that happens to be riskier."*

Kraft's personal opinions on some of Apollo-Saturn's builders I'd met were forthright. He did not hesitate to rate Dr. Gilruth way up there in a class with a Von Kàrmàn or Teller.

"How about Von Braun?"

"Good man. But not in it with Bob Gilruth as an engineer," he said without hesitation. (Huntsville and Houston are, of course, in "rival" camps.) We discussed Storms and Shea.

"I respect Stormy, but he's explosive inside—not the easiest man to talk to—like trying to grab a drink of water from a firehose. Both men were doing a helluva job and were probably resigned to the familiar fate of eventually being replaced. Joe Shea, and I don't mean this in an uncomplimentary sense, was handicapped by trying to carry too much of the burden on his own shoulders. The load can get too big for one man, and Joe needed to rely more on others."

An associate of Shea's, on the other hand, has told me that Joe relied too *much* on the assurances, for instance, of contractors. I then asked Kraft about the Mascon problem—spots on the Moon where concentrations of abnormal density, hence gravity, had affected calculations on Apollo 10.

"I think Mascons have been overplayed in the press," he said. "Gene together with the crew can adjust okay to deviations in altitude, I'm sure, and lateral variations."

I left Mission Control with an illusion of utter unreality. Were these unassuming specialists I'd just talked to fully aware of the part they were playing in one of man's

*Kranz was to make many such tradeoffs during the ensuing eight days.

wildest dreams, in which they were beating the science fiction writers at their own game? Oh well, seeing would be believing—nothing to it, actually—after which we could all go about our business as usual.

Trying with indifferent results to slide off to sleep that night, I wondered how many dozens, hundreds, or maybe thousands of the unsung builders of Apollo-Saturn, with the showdown imminent after all the years of doubt, were having similar difficulties.

ii

Comfortably ensconced at an early hour next morning before a pair of large color television sets provided for the media, along with coffee and samples of the frozen Stouffer food catered to Armstrong's crew, I found myself superimposing on the Stack the faces of its builders, as the camera held on the motionless monolith for long periods. In the weeks of my travels since Apollo 10, as I became familiarized with many of the cast, the sight of the Stack had subtly changed from an impersonal to a very personal involvement. It helped to know the names and numbers of the players.

Beginning at the top, my eye and mind lingered long on each segment.

The world of "Irv" Spitzer and his Lockheed crew, and of the LES men at North American Rockwell, would be centered on the Launch Escape System, praying that it would not be needed, but mindful of the ever-present possibility that Armstrong, Aldrin, and Collins might suddenly find it the most important part of the whole Stack.

The world of Tom Jones, president of Northrop, and of his parachute wizards Steyer, Freeman, and Knacke, would likewise be centered on the possibility of an abort on the pad or during climb-out, hence the crucial role of those parachutes, crushed into the density of hardwood in a narrow collar below the escape tower; and they would be sweating out the dual role of those nylon canopies for eight days, until all the chips would be down on a safe Earth Recovery System.

Many faces drifted across the tiny nose of the Stack, quarters for the crew, and the only segment scheduled to survive the long journey at splashdown: Faget and Allen, proponents of its blunt-end conical shape that thwarted an incinerating reentry; Lee Atwood, master builder, now vindicated by the prior events of four Apollo manned Space flights; Joe Shea and Stormy Storms, who ran the early, uphill laps, before passing on the torch; at Houston, Gilruth, Low, and their teammates; at Downey, Bergen, Myers, Ryker, Feltz, Healey, Jeffs, Fields, among many more; at Cambridge, Massachusetts, "Doc" Draper the navigator, and Dr. James Elms, director of NASA's Electronic Research Center; for the exasperatingly exotic environmental control system, Garrett's boss Harry Wetzel, his program manager Carl Jackson, engineer Henry Nicolello, and biomedical expert Dr. James N. Waggoner; for the heatshield's ablation technology, Avco's Dr. Clifford Berninger, Ed Offenhartz, and Walter Zeh; for the life-supporting "backpack" for the astronauts' spacesuits, the environmental control system of the lunar lander and its abort-sensing gear, the team at Hamilton Standard division of United Aircraft; for the biomedical well-being of the men in the cabin, MSC's Dr. Charles A. Berry. For all of these men, and colleagues at Honeywell, Kollsman Instruments, Raytheon, GM-AC, and GE concerned with the wigwam, time was about to stand still.

At the next lower level, the Service Module, the world for Mike Vucelic and Bob Fields would center on the wigwam's flying "supply depot" and its SPS propulsion system, and the worlds of Herman Coplen and Dan David would center on the 101 percent reliability they had tried to implant in their Aerojet engine for the SPS.

Below the Service Module, where the LM bug reposed, the man who built its "garage"—the Adapter Section—at Tulsa division of North American Rockwell, Harry W. Todd, and manager Charlie Halstead, would be diverting their attention besides to the segments above and below the Lunar Adapter section, respectively, to the primary structure of the Service Module and to a considerable portion of the S-II stage (by weight), both of which had been built at Tulsa, plus the primary structure of the IU (Instrument Unit, the brain of the Saturn-V booster, built by IBM).

The invisible lunar bug, soon to be the cynosure of world attention, was blotting out all else for Carroll H. Bolender, NASA's program manager for the LM, and Joe Gavin and Tom Kelly of Grumman, not to mention TRW's Gerard W. Elverum, Jr., builder of the Bug's throttleable Descent engine. Elverum, of course, shared with Aerojet the fearful obligation of having to build another of the three engines that must not fail under any circumstances, period, the third "must" being saddled on the builders of the Ascent engine for lunar departure, a joint responsibility of Rocketdyne and Bell Aircraft Company. Also from TRW, builder Dave Meginnity would be losing sleep awaiting a successful performance by the LM's guidance system, complete in itself but serving also as a backup to Draper's primary system in the Command Module.

My gaze shifted a notch lower, to where the "payload" ended, measuring less than one-fourth of the total Stack, down to the top of the three stages of the Saturn V, capped by the ring of IBM's Instrument Unit. One brooding face overshadowed all others, unforgettable, that of booster-builder Wernher Von Braun, suffusing the whole rocket "carrier," with co-artisans Oswald Lange and Eberhard Rees looking on against the ghost of Dr. Robert H. Goddard.

Then, upon the topmost S-IVB third stage—the stage that must provide Apollo with its final shove up to Earth-orbital speed and later to trans-lunar speed from its single J-2 engine—came through the faces of designers Patelski and Thiele, and builder T. D. Smith, of McDonnell Douglas' division at Huntington Beach, California.

Bob Greer hovered over the much larger S-II second stage, sitting on the five-engine cluster of Castenholz's J-2's, both of the troubled history. And finally, Boeing's highrise of a first stage, the champion heavyweight lifter that must shoulder the initial brunt of the entire 6.5 million-pound vehicle, stared back at me with the features of Boeing's William Allen, and of bantamweight Sam Hoffman of Rocketdyne, whose life efforts had culminated in making it possible to hoist the Stack exactly one inch—that all-encompassing first one.

Looming large overall in the background, like profiles on a cosmic Mt. Rushmore, at the elevation of high decision making, were the faces of Johnson, Kennedy, Glennan, Webb, Dryden, Holmes, Seamans, Paine, Houbolt, Mueller, and last but not least Phillips.

Once more the moment came. Your heart stopped as the countdown reached "T minus ten seconds." Once more it was as if the Empire State Building had risen from the pavement of Manhattan, and Apollo 11 was on its way.

I breathed easier for Von Braun and Lee Atwood and Al Rockwell after the first and second booster stages had performed their duty and dropped away into the Atlantic Ocean. I breathed easier for K. J. Katelski and T. D. Smith after the third stage had accelerated the spacecraft into a good Earth orbit, fired on schedule to catapult the crew out of orbit and into Trans-Lunar Injection, and likewise separated and plummeted. Now, for the next quarter of a million miles, in an extremely close approximation of a perpetual motion machine, burning no fuel, and of an antigravity machine, Apollo 11 was embarked on a free ride. Its builders could only wait. And wait.

I caught the afternoon plane for home and a ringside seat with my family. Every American with access to a TV set now had as good a view as Gilruth at Mission Control, four days later, when TRW's descent engine fired and Armstrong began his historic approach down toward Tranquillity Base. But I was thinking not so much of

Armstrong and what his pulse rate must be, as of the systolic blood pressure of one Gerard Elverum, who had confided to me his greatest fear pending an actual manned lunar landing.

What he could not predict, in the absence of direct experience with the Moon's crust, as the hovering Bug lowered itself the last few feet atop the thrusting flame of the descent engine, was whether there might occur a back pressure that would "balloon" the Bug upward in a sort of bounce from the surface. Worse still, he could not be sure that the rocket's flame would not deflect upward from that alien surface into the engine's innards and burn up vital accessories or cause other serious damage.

The first eventuality, if repetitious, could cause Armstrong to deplete all his fuel before a safe landing was assured, and the second eventuality might be equally disastrous. When neither happened, and the Bug touched down to what Elverum later told me was "too good to expect, a perfect landing," I inwardly raised a toast to him. But I felt no gratitude for the inside information by which he had placed an overload on my nerves during a climax of suspense that was already downright unbearable.

iii

When an emotion-choked President Nixon congratulated Armstrong, Aldrin, and Collins through the window of their quarantine trailer aboard the recovery carrier Hornet, and the leaders and citizens of all nations echoed his sentiments, few people anticipated how soon a much longer visit on the Moon, Apollo 12, would evoke queries of, "So what? When you've seen it once, you've seen it all."

One charming gentleman, short of stature but long on a sense of humor, with the high-bridged nose of the executive that often goes with a good commander, did not share this view—Apollo 12's skipper Charles "Pete" Conrad. I met Pete a few weeks before his mission, scheduled for November 14.

"Neil Armstrong is not an easy act to follow," he acknowledged.

He did not say so, but the tenor of his ensuing remarks was edged with awareness that his "act" would in some ways be tougher. Armstrong had raised the curtain on part of the lunar unknowns, but there would be other unknowns for Conrad. While Richard F. Gordon circled overhead in the Command ship, he and Alan L. Bean would explore much further out from their haven in the LM, providing more stringent tests of the handiwork of the men at ILC Industries who built their spacesuits, in their performance of new tasks in setting up the equipment for scientific experiments and to bore deeper into the lunar skin for samples.

The toughest part of Conrad's act was actually going to occur a short distance above the palm trees of Florida, when a bolt of lightning turned on every red bulb in the cabin—a golden opportunity for any but the coolest of operators to press a lot of wrong buttons—or decide that the "Abandon Ship" button was the right one.

This was still in the future, however, and I turned the conversation to a lighter vein. Pete and his animatedly lovely wife Jane, taller than he in inches but not when it came to wearing the pants, were relaxing over cocktails at the home of Conrad's friend "Bud" Joyce. The latter is president of Dyna-Therm Corporation of Los Angeles, which deserves passing mention because it typifies the contributions of small business (which space has constrained me to neglect) by having come up with the flame-resistant coatings which NASA employs at its Apollo launch complexes to protect vulnerable equipment.

"A lot of people have been curious," I prodded Conrad, "just how long astronauts have to wait for a belt of booze after the strain of a Space mission. Dr. Waggoner, the flight surgeon, solemnly assured me that your quarantine trailer will be strictly 'dry' for days after splashdown. Isn't that kind of rough?"

"All I can say is," he said, rotating a frosty martini

glass in his fingers, "that once the preliminary physical exam on the carrier is over, I know more than one astronaut who's going to raise unshirted hell if that jug of martinis isn't waiting there for him. Well chilled."

Surely Conrad had richly earned such a reward at the hands of an enlightened medico, after he and Bean and Gordon had boarded their recovery chopper on November 24 to climax a virtually "nominal" Apollo 12 mission.

A portion of the taxpaying electorate, and of the Space-satiated teenage audience, audibly yawned. But Apollo 13, laid on for April 13, 1970, with liftoff to take place at 1313 hours, was waiting offstage to stifle the yawns.

iv

Chris Kraft, a man not prone to exaggeration, has stated that the best thing that could have happened in a frighteningly bad situation was Gene Kranz's happening to be on duty at Mission Control when an explosion in the Service Module crippled commander James A. Lovell's spacecraft far out en route to the Moon. He added that it was a good thing, too, that Kranz was on the job again for the critical final phase of the Apollo 13 cliffhanger.

There being few readers anywhere not familiar with the epic struggle of Jim Lovell, Fred W. Haise, Jr., and John L. Swigert, Jr., to survive the prolonged catastrophic emergency that befell them, and this being a book about builders rather than Space missions, I will dwell only briefly on the operational side of things. Gene Kranz's performance illustrates both of Chris Kraft's points about black magic and about tradeoffs.

Intuition had to lie behind Kranz's surprising decision to allow a fresh team of controllers go on duty as scheduled, thereby changing horses in midstream at the most awkward of junctures early in the emergency. And a sure instinct lay behind the irrevocable early decision to go the "long way home," all the way around the Moon and back, instead of surrendering to the more natural urge to fetch the astronauts back to Earth by speedier, if riskier, means.

Gene, for his part, has voiced unstinted praise not only to the crew, but to the contractors' representatives at his elbow, so to speak, who helped him to provide the many ingenious improvisations that enabled the astronauts to make do, and to the creators of the extraordinarily versatile lifeboat, the beautifully ugly little bug that Grumman built.

North American Rockwell, builders of the Service Module in which a malfunction had deprived the crew of its primary source of electrical power and life support systems set about the correction of a freak of misfortune which had not occurred in six prior Apollo missions.

The findings of a NASA review board, released in June 1970, revealed that the explosion of one of the Service Module's two oxygen tanks was the culmination of an "unforgiving" design and a series of mistakes, committed long before Apollo's blastoff, by the prime contractor, by subcontractor Beech Aircraft as supplier of the oxygen tanks, and by NASA personnel, the blame being shared by all three.

In an earlier statement, NASA had announced a delay for Apollo 14's launch date, pending the correction (and it was confident of positive correction) of the cited discrepancies. Root source of the accident, NASA disclosed, was identical with that in the fatal tragedy on Pad 34 in 1967: the proximity of combustible materials to a source of ignition in an atmosphere of 100 percent oxygen under high pressure, with "ignition" resulting from an electrical arc after a wiring short circuit (presumably to a fan motor).

The parallel did not escape editorial critics, who demanded to know, in effect, "When will we learn?" And as in 1967, in effect, there were cries that "somebody ought to be hung."

The question was legitimate. If the same mistake had

indeed been repeated, why? With the jury still out on the intricacies of similarity in potential fire hazards, I can only submit that in any case we *have* learned, in both instances.

We have been reminded once more, or should have been, that despite the spending of billions of dollars to assure ultra-reliable hardware, and despite the most prodigious efforts of the most intelligent and dedicated men that can be found, the end product, be it a jet transport or a spacecraft, is no better than the combined result of men *relying* upon each other. The vaster the enterprise, the more precarious the interreliance. The greater the complexity, as in Space, the graver the consequences of even the lowliest error.

It is a cliché, as well as the sad truth, that men always have made, always will make, mistakes. Less obvious is the wide scope of the mistakes that have been committed in every aircraft, missile, and Space program yet undertaken by any nation. If the public is under an illusion that mistakes in a project like Apollo are and should be rarities, it might as well realize that the records of NASA and every contractor will show that there have occurred millions, probably billions, of individual mistakes, most often occurring when one worker has over-relied on the work of another. Accordingly the very guts of the Apollo-Saturn program has centered on taking the measures between 1961 and 1970 to detect those millions or billions of errors, to correct them in time, and to forestall their recurrence, in a quest for 99.99 percent reliability.

When something has gone grievously wrong, as on Apollo 13, it was because errors had squeaked through between that last decimal and the ideal of 100 percent, in some area where one or more men relied once too often on another's work. The strident critics want the perfect 100 percent. They will never get it. What they *have* gotten, however, is a record of accomplishment by our government-industry team on Mercury, Gemini, and Apollo, during which not a single American has died in

Space, that has exceeded the most sanguine expectations and defied belief itself.

It is a record that deserves the gratitude and admiration of all of us who still believe that, far from overplaying our hand in the national effort to begin exploration of our universe, we have only scratched the surface of its potential benefit for the human race. For those not interested in "potential" benefits, tangible benefits in abundance are here right now for the doctor, for the weather forecaster, in agricultural benefits, in the management and exploration of natural resource, in communications, and in many more areas. But the message, somehow, doesn't seem to be getting through to the customer. Let me cite just one simple example under the heading of entertainment, or recreation. Eight hundred million viewers were reported to have watched the 1970 World Soccer Championships in Mexico on TV, thanks to a global broadcast via communications satellite. That's a billion dollars a day's worth of entertainment, assuming that each viewer would have been willing to pay the price of a movie, say $1.50. When you add this up for the several days of the championship, and include other sports events bounced off satellites for an international audience, like Wimbledon, the British Open, the World Series, or a heavyweight championship fight, during a given year, you've just about paid for the entire Apollo-Saturn program in the value of entertainment dollars alone.

V

Are the major aerospace firms becoming discouraged by hot-and-cold public reaction to the budgeting of NASA's projections for post-Apollo Space shuttles, orbiting workshops, and visits to more distant celestial bodies, on which industry's planning must be based? Or by the venom of a widespread campaign in some of the news media to portray the government-industrial complex as a plot to aggrandize the supposedly swollen profits of the very companies who have been losing their shirts on aerospace contracts to the verge of bankruptcy, and who have been straining now for years to diversify into more profitable commercial business?

For representative answers, I consulted Robert Anderson, president of North American Rockwell, which is still the leading contractor in terms of Space dollars, though no longer first across the spectrum of aerospace commitments.

"Would you comment," I asked Mr. Anderson, "on certain outside rumors since the merger, that the Rockwells never intended to emphasize North American's future in aerospace, but planned rather to cash in on its capital assets for the benefit of the Pittsburgh side of the house?"

"I certainly would," he said, "having heard the same implications, as you say, that we were out to deemphasize North American's traditional programs. Well, I'll let the record speak for itself. First, we went into the merger with our eyes open to a heavy immediate loss over at Space Division for spacecraft modifications after the fire, and related losses on canceled incentive payments; then we invested about $25 million in a bid for the F-15 fighter plane contract, took our lumps on that one, following which we bet another $25 million, approximately, on the B-1 bomber contract, which we won on the best design. Through all of that, we were continuing our support of Space Division's efforts to win its share of post-Apollo work.

"At the same time," he continued, "the B-1 is probably the last of the big military airplane contracts. And there is no immediate prospect of a new Space contract as big as Apollo, although we are not ruling out the possibility of even more important Space ventures, as an outgrowth of the very considerable post-Apollo developments now under way.

"Those are hard facts that you can't wish away. Even if there had been no merger, and say North American had *not* won the B-1 competition, its management would have been faced with exactly the same unpleasant alternatives as we now face together. You'd simply have to take whatever measures were necessary for self-preservation. As it stands, we're as determined as anyone to excel in the field of aerospace, and to place bets on its future."

vi

For Colonel Rockwell and his son Al, North American Rockwell's driving chairman of the board, history may well be repeating itself. Chronically infected with confidence in the brightness of this country's industrial future, the Colonel bucked the trend 40 years ago at the onset of the Great Depression by starting to build a Pittsburgh-based manufacturing empire that survived and flourished. The son learned well at his father's knee.

Bad as the timing of the elder Rockwell looked at that ominous juncture, it appeared to many in the financial community that the timing was even worse when the father-son team chose to enter the field of aerospace, teamed with North American, at the very time that a depression was setting in for the industry, with government spending for both Space and military aviation shrinking rapidly, and after funding for our ballistic missile force was past its hump.

Those who know Al Rockwell, Jr., well, suspect that once again the Rockwells have embarked on a course that will confound the skeptics. Al's is not the stuff of which losers, or quitters, are made. Nor is Anderson's. Both are keenly aware that in technology, "You ain't heard nothin' yet," in the imperishable prose of Al Jolson. And both are looking ahead, for example, to felicitous commercial payoffs from aerospace "technology transfer" and to the eventual fruition of some of the visionary planning going on over at Bill Bergen's Space Division.*

*Bergen has been promoted and succeeded by Joseph P. McNamara as president.

Kraft Ehricke at Space Division envisions a world where, for an economical fare, American tourists will be able to shuttle up to a 1,200-room hotel in Earth orbit, offering artificial gravity and all of the luxuries of the Hilton chain—in the not-so-distant future, too; experimental Space shuttle vehicles are well along on the drawing boards, and NASA is already developing an orbiting workshop at Huntsville, using an adaptation of McDonnell Douglas' S-IVB third Saturn stage. Ehricke has the costs for his hotel figured right down to the weekend rate for a couple and the price of a gourmet dinner. He foresees, also, great promise in manufacturing carried on in near-Space, where weightlessness and a perfect vacuum offer enormous advantages, and where certain processes impossible on Earth become relatively easy.

The all-pervasive battle against gravity, he will remind you, accounts for most of the expense of building and operating our factories—witness, for example, all those huge, costly overhead cranes, not needed in Space to lift an object that weighs zero.

Other visionaries like Ehricke share with most TV viewers, having observed planet Earth for the first time from another heavenly body, the eerie sensation that we are coming close to piercing the mystery of why men came to exist there, and how.

vii

Nearly 40 years ago it was my privilege and good fortune to join what was then a small group—the company of military pilots. Out of a graduating class of 91 flying Cadets at Kelly Field, Texas, in 1933 (which inluded Benny Schriever), my aptitude as a pilot probably placed me in the bottom half dozen. I say "probably" because I've never seen my grades, but I do know that in the ensuing years on active duty in the Army Air Corps, which evolved into the United States Air Force, I was marginal enough as a pilot to feel like something of an imposter among my more talented peers. But then a rewarding and unlikely role began to open up for me. I became one of the early spokesmen for those men, trying to communicate to readers the rare experience which we shared, and which my compatriots could not, or at least did not, express in words.

Over those years, I have had the gratifying experience of having strange fliers of all ages and degrees of distinction come up to me and place the blame squarely on my book, "I Wanted Wings," for having gotten them into the flying business. Astronauts, too, have been among them, and Space managers like "Chuck" Allen, in charge of rocket test-firings at the Mississippi Test Facility, Bay St. Louis.

Flying, of course, has long ago become commonplace. But Space flight has not, by any means. The force that has driven me to write these pages about the builders of Apollo-Saturn is the same one that drove me to write about aviators—naked, unashamed admiration.

I apologize to no one, if what I have seen of the "shoulders" of Apollo's "giants" has biased me on their behalf. But I apologize to all if I have been derelict in my effort to do them the justice that they, like our pioneer pilots, deserve.

The reward that men prize as highly as any other, if not higher, is recognition, the approbation of their fellow men. For when all is said and done, the sweetest music this side of heaven, usually heard sparingly, grudgingly, too late, or not at all, is the phrase coming from an athletic coach to a boy, from a college mentor, from a boss, or even from an author:

"You did a good job."

That truly big little man, Dr. Charles Stark Draper, late of M.I.T., in an understatement that came from the heart, spoke for all the builders of the Stack, when he said, as nearly as I can recall his words:

"The pleasure of seeing some new system of yours work well is surpassingly great."

NASA AWARD WINNERS
(As of December 31, 1969)

DSM *(Distinguished Service Medals)*

Anders, William A.
Bisplinghoff, Raymond L.
Bogart, Frank A.
Bolender, Carroll H.
Borman, Frank
Bourdeau, Robert E.
Buckley, Edmond C.
Carpenter, M. Scott
Cernan, Eugene A.
Chaffee, Roger B.
Clark, John F.
Clark, Raymond L.
Cooper, Percy G.
Cortright, Edgar M.
Covington, Ozro M.
Crowley, John W.
Debus, Kurt H.
Dembling, Paul G.
Dryden, Hugh L.
Faget, Maxime A.
Flax, Alexander H.
Gilruth, Robert R.
Glenn, John H.
Glennan, T. Keith
Gorman, Harry H.
Grissom, Virgil I.
Gruene, Hans F.
Hage, George H.
Hjornevik, Wesley J.
James, Lee B.
Jones, David M.
Kleinknecht, Kenneth S.
Kraft, Christopher C.
Lovell, James A.
Low, George M.
Mathews, Charles W.
McDivitt, James A.
Mitchell, Jessie L.
Mueller, George E.
Naugle, John E.
Newell, Homer E.
O'Connor, Edmund F.
Petersen, Forrest S.
Petrone, Rocco A.
Phillips, Samuel C.
Pickering, William H.
Purcell, Joseph
Rees, Eberhard F. M.
Richard, Ludie G.
Rudolph, Arthur L. H.
Scheer, Julian W.
Schirra, Walter M.
Schneider, William C.
Schriever, Bernard A.
Schweickart, Russell L.
Scott, David R.
Seamans, Robert C.
Shapley, Willis H.
Shepard, Alan B.
Siepert, Albert F.

Silverstein, Abe
Slayton, Donald K.
Sohier, Walter D.
Stafford, Thomas P.
Thompson, Floyd L.
Truszynski, Gerald M.
Von Braun, Wernher
Walker, Joseph A.
Webb, James E.
Weidner, Hermann K.
White, Edward H.
Williams, John J.
Williams, Walter C.
Young, John W.

Draper, Charles S. (DPSM)

ESM *(Exceptional Service Medal)**

Algranti, Joseph S.
Alibrando, Alfred P.
Aller, Robert O.
Amman, Ernest A.
Arabian, Donald D.
Armstrong, Neil A.
Artley, Gordon E.
Ashworth, C. Dixon
Atkins, John R.
Auter, Henry F.
Bakutis, Fred E.
Balch, Jackson M.
Bales, Stephen G.
Ballinger, Edward P.
Barnett, Henry C.
Baron, Oakley W.
Bakron, Paul A.
Bass, William P.
Baumann, Robert C.
Bavely, James C.
Belew, Leland F.
Bell, Lucian B.
Bellman, Donald R.
Bernardo, James V.
Berry, Charles A.
Bertram, Emil P.
Bethay, Joseph A.
Bland, William M.
Bobik, Joseph M.
Bolger, Philip H.
Bond, Aleck C.
Boone, Walter F.
Bowman, Julian H.
Boyer, William J.
Bradford, James E.
Bramlet, James B.
Brinkmann, John R.
Brock, Eugene H.
Brockett, H. R.
Brooks, Charles O.
Brown, B. Porter
Brown, William D.
Brownstein, Herbert S.
Bruns, Rudolph H.
Buchanan, Donald O.
Buckley, Charles L.
Buckner, Garland S.
Burcher, Eugene S.
Burdett, Gerald L.
Butler, Herbert I.
Calio, Anthony J.
Call, Dale W.
Cariski, Sidney A.
Casani, John R.
Cataldo, Charles E.
Catterson, Allen D.
Chamberlin, James A.
Charlesworth, Clifford E.

*Holders of DSM also not repeated here.

Chauvin, Clarence A.
Cheatham, Donald C.
Chilton, Robert G.
Clarke, Victor C.
Cohen, Aaron
Cohen, William
Collins, Michael
Condon, John E.
Conrad, Charles
Constan, George N.
Cook, Richard W.
Cooper, Leroy S.
Crane, Robert M.
Cunningham, Ronnie W.
Dannenberg, Konrad K.
Darcey, Robert J.
Davies, M. Helen
Day, LeRoy E.
Deutsch, George C.
Dineen, Richard C.
Donely, Philip
Donlan, Charles J.
Donnelly, Paul C.
Driscoll, Daniel H.
Duerr, Freidrich
Duff, Brian M.
Duncan, Robert C.
Dunseith, Lynwood C.
Easter, William B.
Edwards, Marion D.
Eisele, Donn F.
Eisenhardt, Otto K.
Elms, James C.
Fannin, Lionel E.
Felberg, Fred H.
Fichtner, Hans J.
Foster, Joyce N.
Foxworthy, Davis E.
Frasier, Cline W.
Frutkin, Arnold W.
Fuhrmeister, Paul F.
Funkhouser, James M.
Gabriel, David S.
Garbarini, Robert F.
Gardiner, Robert A.
Garrick, I. Edward
Gaskins, Roger B.
Gaver, Austin L.
Gay, Clarence C.
Geissler, Ernst D.
Gibbons, Howard I.
Ginter, Roll D.
Godfrey, Roy E.
Goerner, Erich E.
Goldcamp, Thomas F.
Gordon, Richard F.
Gorman, Robert E.
Grau, Dieter
Gray, Robert H.
Gray, Wilbur H.
Greenglass, Bert
Guthrie, Crompton A.
Hafussermann, Walter
Haglund, Howard H.
Hagood, Carlos C.
Haley, Richard L.
Hall, Charles F.
Hamilton, Harry H.
Hammack, Jerome B.
Hammers, Fred C.
Hardeman, Theodore U.
Harris, Gordon L.
Hawkins, Willard R.
Hearth, Donald P.
Heimburg, Karl L.
Heiser, Robert F.
Heldenfels, Richard R.
Hicks, Ralph L.
Hill, Paul R.
Himmel, Seymour C.
Hirsch, Oliver M.
Hobokan, Andrew
Hodge, John D.
Hodgson, Alfred S.

Hoelzer, Helmut
Hoffman, Robert W.
Holcomb, John K.
Hood, Arthur F.
Hosenball, S. Neil
Hueter, Hans
Hursey, Benjamin W.
Huss, Carl R.
Huston, Vincent G.
Huth, Chauncey W.
Isbell, Thomas P.
Jaffe, Leonard
Janokaitis, John J.
Jean, Otha C.
Jenkins, Thomas E.
Johnson, David S.
Johnson, William G.
Johnson, Bernard L.
Johnson, Caldwell C.
Johnson, Marshall S.
Johnson, Robert E.
Johnston, Richard S.
Jonash, Edmund R.
Kelleher, John J.
Keller, Samuel W.
Kelley, Albert J.
Kemmerer, Walter W.
King, Charles H.
King, John W.
King, Robert E.
Kingsbury, James E.
Kinzler, Jack A.
Kotanchik, Joseph N.
Kramer, James J.
Kranz, Eugene F.
Kroeger, Hermann W.
Kroll, Gustave A.
Krueger, Donald A.
Kubat, Jerald R.
Kyle, Howard C.
Lagow, Herman E.
LaHatte, William F.
Lang, Dave W.
Lealman, Roy E.
Lederer, Jerome F.
Lee, Chester M.
Lesher, Richard L.
Lilly, William E.
Lindberg, James P.
Lloyd, Oakley B.
Loftin, Laurence K.
Lohse, William M.
Long, Robert G.
Lubarsky, Bernard
Lucas, William R.
Ludwig, George H.
Luedecke, Alvin R.
Lunney, Glynn S.
Lynn, H. Robert
Mack, Jerry
Mahon, Joseph S.
Malaga, Joseph F.
Mandel, Carl H.
Marcotte, Paul G.
Mars, Charles B.
Marshall, William R.
Martin, James S.
Mathews, Edward R.
Maus, Hans H.
Mayer, John P.
Maynard, Owen E.
McCaffery, Robert J.
McClanahan, Jack T.
McClintock, John G.
McCool, Alexander A.
McCulloch, James C.
McGuire, Charles W.
McLane, James C.
McMullen, Thomas H.
Mengel, John T.
Middleton, Roderick O.
Miller, Frederic H.
Miller, William E.
Milwitzky, Benjamin

Minderman, Peter A.
Mohlere, Edwar D.
Moore, Fletcher B.
Morea, Saverio F.
Morgan, Homer G.
Moritz, Bernard
Morris, Mildred V.
Morris, Owen G.
Moser, Robert E.
Mrazek, William A.
Murphy, James T.
Myers, Boyd C.
Nelson, Clifford H.
Nettles, J. Cary
Neubert, Erich W.
New, John C.
Newby, David H.
Newman, Charles T.
Nichols, Steward H.
Nicks, Oran W.
Noel, George W.
North, Warren J.
O'Brien, Gerald D.
O'Hara, Alfred D.
Olson, Royce G.
Oswald, Donald R.
Owen, Robert L.
Page, George F.
Parker, Clarence C.
Parks, Robert J.
Parry, Edward F.
Parson, John F.
Paul, Henry C.
Pickering, John E.
Pickett, Andrew J.
Pierce, Edward A.
Piland, Robert O.
Plohr, H. Warren
Potate, John S.
Powell, James T.
Pozinsky, Norman
Press, Harry
Preston, G. Merritt
Price, Paul A.
Rabey, Duncan W.
Rafel, Norman
Raines, Martin L.
Rainey, A. Gerald
Rainwater, Wallis C.
Rapp, Robert A.
Ream, Harold S.
Reed, Robert D.
Reinartz, Stanley R.
Renzetti, Nicholas A.
Reyes, Raul E.
Rieke, William B.
Rigell, Isom A.
Ritland, O. J.
Roberts, Tecwyn
Robinson, Lorne M.
Ross, Miles
Rushworth, Robert A.
Sargent, Jack
Sasseen, George T.
Sasser, James H.
Savage, Melvyn
Sawyer, Ralph S.
Schaibley, John R.
Scherer, Lee R.
Schick, William H.
Schindler, William R.
Schmittling, Donald L.
Schurmeier, Harris M.
Scull, Wilfred E.
Sendler, Karl
Shapiro, Ralph
Shea, John
Shea, William M.
Sheperd, James T.
Simmons, William K.
Simpkinson, Scott H.
Simpson, George L.
Sjoberg, Sigurd A.
Skaggs, James B.

Slattery, Bart J.
Small, John W.
Smith, Spencer E.
Smith, Richard G.
Smylie, Robert S.
Sneed, Bill H.
Sorensen, Victor C.
Speer, Fridthof A. H.
St. Clair, Charles W.
Stanley, Hubert R.
Stelter, Laverne R.
Sterett, James B.
Stevenson, John D.
Stimson, Bailey E.
Stoney, William E.
Suhlinger, Ernst
Styles, Paul L.
Taylor, Eldon D.
Teir, William
Tepper, Morkis
Tessman, Bernard R.
Thibodaux, Joseph G.
Thompson, Henry F.
Thompson, Robert F.
Thomson, Jerry
Tindall, Howard W.
Tolleson, Robert T.
Trimble, George S.
Trott, Jack
Urlaub, Mathew W.
Vaccaro, Michael J.
Van Staden, George A.
Varson, William P.
Vaughan, William W.
Vavra, Paul H.
Vecchietti, George J.
Vette, James I.
Vreuls, Frederick E.
Walton, Thomas S.
Wasielewski, Eugene W.
Weiland, Stanley
Whitbeck, Philip H.
White, George C.
Wiple, M. Keith
Widick, Herman K.
Wilkinson, Reuben L.
Williams, Francis L.
Williams, Grady F.
Wilson, Herbert A.
Wood, H. William
Wood, Roy E.
Wood, William H.